30 INNOVATIONS

OF THE RUSSIAN ENGINEER

ANATOLY ROZENBLAT

ISBN: 1-4033-5637-8 (e-book)
ISBN: 1-4033-5638-6 (Paperback)

This book is printed on acid free paper.

1stBooks – rev. 06/03/03

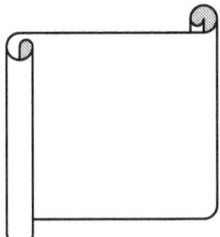

Dedicated with love, to my dear father,
Rozenblat Isaac Samuelovich

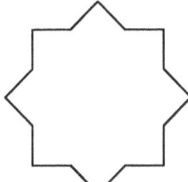

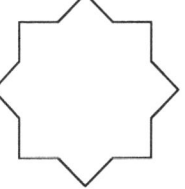

TABLE OF CONTENTS

APPENDICES

PREFACE

The technical progress of any society leads not only to the increasing of complicated technology and production, but at the same time forces everyday to search the new views of production.

And of course, this demands a lot of time and investments on designing and testing of the different views of production for each manufacturing plant.

The designing of new technique puts before of any designer the definite tasks and problems in question of the novelty of this object and minimum cost for the production of this device with the next evaluation of its technico-economical indexes and terms of compensation.

This book gives to any designer some useful reference information as to make the right decision in selection of suggested by author innovations at this book and how to evaluate its efficiency for production I industry.

At the same time this could help the designer to reduce the term of designing the new project and also the investment on the development of this new device.

All book almost is devoted of discovering the new innovations which were designed and described by author in the first approaching as the novelty and of course demands further detail investigation in question of fulfilling of the experimental-research works and definite investments for production of this device.

But the worth of this information which is given at this book is that any designer could use the innovations which are suggested by author concretely to his manufacturing plant with the next evaluation of the technico-economical indexes of this designed device.

This book gives a brief review of inventions with special attention to the advanced machining technology for commercial practice and further research in the different manufacturing processes in the 21 st Century.

Specifically, there are four types of inventions discussed here: *environmental, safety of flying, advanced machining processes and cutting tools, the efficiency of some machines and mechanisms is also discussed.*

In addition, this book indicates basic design and research activities of author in period of 1990 to 2001 in the United States of America.

June 30,2002

Anatoly I.Rozenblat
Independent Scientist and Inventor,
Member of ASME and SNAME

Chicago, USA

CHAPTER ONE
BASIC DESCRIPTION OF THE INVENTIONS

A. PATENTS IN FORMER USSR

1.Combination lathe tool of A.I.Rozenblat

SPECIFICATION TO AUTHOR'S CERTIFICATE

UNION OF SOVIET SOCIALIST REPUBLIC
(19) SU (11) **1220856 A**
(51) 4B 23 B 27/00

USSR STATE COMMITTEE ON INVENTIONS AND DISCOVERIES

(21) 3760952/25-08
(22) 07.06.84
(46) 03.30.86 Bull.№12
(72) A.I.Rozenblat
(53) 621.9.025 (088.8)
(56) SOT map, p.5,18,1934

(54) COMBINATION LATHE TOOL OF A.I. ROZENBLAT

(57)   A combination lathe tool which comprises a holder, a cutting part with a head and a stem installed in the holder on a support with the possibility of turning, and a turning mechanism interacting with the stem of the cutting part. The tool is distinguished by the fact that in order to expand its technological capabilities by increasing rigidity of fastening the cutting part, the support is installed on the head of the cutting part of the tool and is made in the form of two convex spherical plates interconnected by a pin with a spherical head.

The invention is related to metal processing. The objective of the invention is to expand technological capabilities of the tool due to expanding the range of adjustment of the front angle of the cutting tool.

Figure 1 shows the suggested combination lathe tool, its section; Figure 2 shows the section A-A from the Figure 1.

A combination lathe tool (1), for example a grooving one, is composed of a plate (2) fastened with the aid of a pin (3) and a stop (4) with a bolt (5), and also a holder (6). On the head of the cutting part 1 is installed a support made in the form of two convex spherical plates (7) and (8), connected through the pin (9) to the spherical head with the aid of nuts(10). The steam of the cutting part 1 of the cutting tool is connected through the crank drive (11), ties (12) and pins (13) and (14) with the holder 6. In order to fix the specified position of the front surface (15) of the cutting tool, on its lateral surface is made a hole (16) into which enters the fixing element (17) in the case of adjusting the cutting part (1) prior to processing a work piece by cutting for the specified constant value of the front angle.

When the front angle has to be changed in the cutting part (1) in the process of cutting, the gear wheel (18) is included which is set into rotation from the kinematics chain of the machine tool, for example from its drive.

The combination lathe tool and the operation during the cutting processing of a work piece are performed in the following manner.

The holder (6) of the combination lathe tool is installed in the tool holder (not shown). The fixing element (17) is removed from the hole (16), the front area (15)of the cutting tool is set and fixed. In addition, the vertex of the plate (2) is matched, for example, with the axis of the work piece. The rigidity of fastening of the cutting part upon cutting and the range of changing the front angle, and also of the cutting angles are ensured due to the fact that the articulated connection has no gaps because of tightening the plate (7) and (8) with by the pin(9).

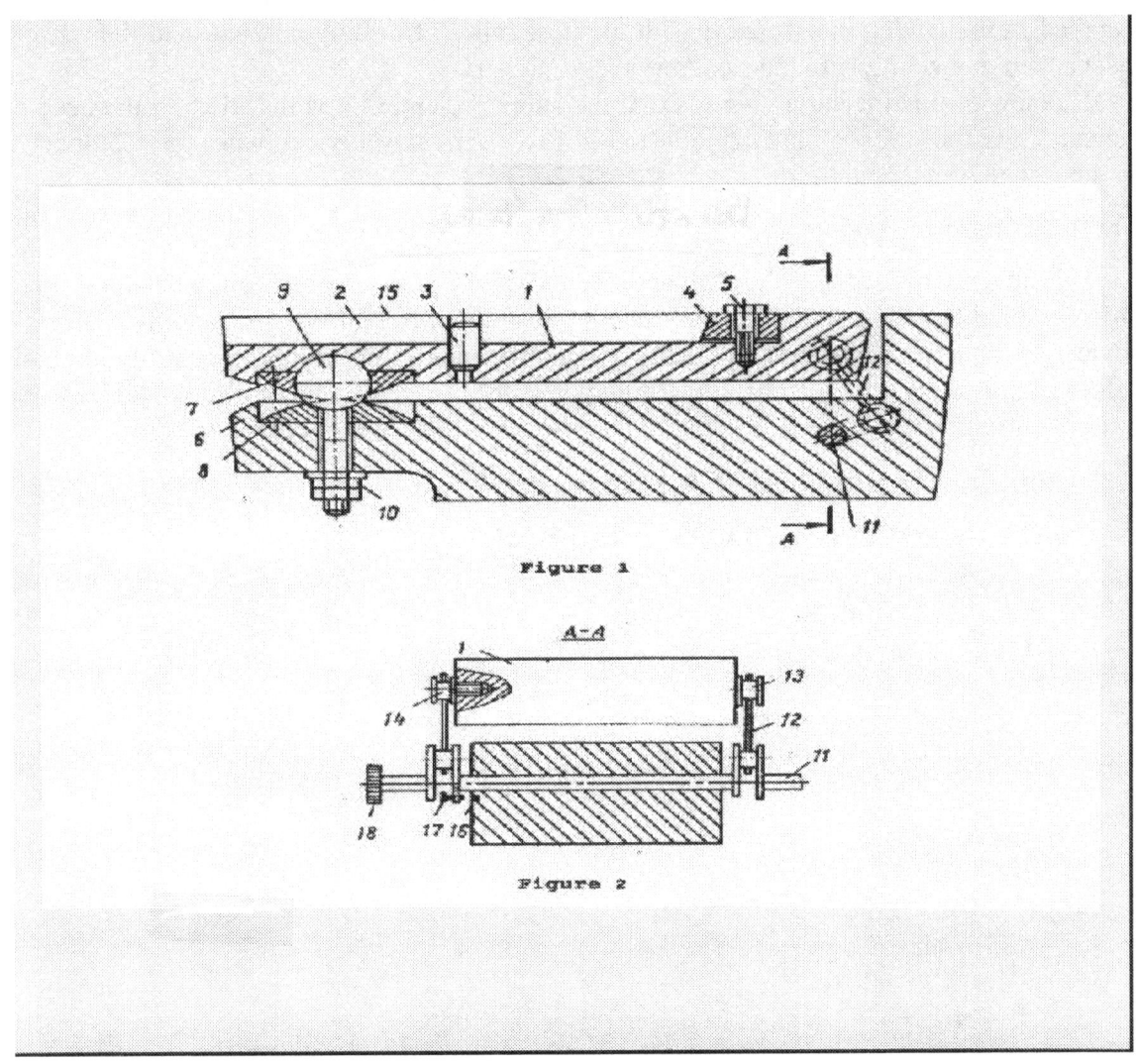

Figure 1

Figure 2

Combination lathe tool

DESCRIPTION OF ADVANTAGES OF THE INVENTION:

"COMBINATION LATHE TOOL OF A.I. ROZENBLAT"

Utility and advantages:

This invention is used to expand the technological possibilities of tool by means of increasing the rigidity of fastening the cutting part of tool and reducing of wear in the process of cutting the different materials advantageously the stainless and high-strength steels. Such materials cause continual wear and failure of the geometry of cutting tool particularly on the face of a tool over which the chip slides forming the chip breaker surface. One of the factor to decrease the wear and increase tool life in during of cutting process is the changing of the side and back rake angles of cutting tool. This invention removes these defects and improves the quality of cutting process.

Novelty:

This invention comprises a holder and also the cutting part with a head, and a stem it installed in the holder on a support with the possibility of turning and has also the turning mechanism interacting with the stem of the cutting part.

The new of this invention there is that the support element it is installed on the head of the cutting part of the tool and made in form of two convex spherical plates intercon-nected by a pin with a spherical head.

Universal and profitable:

This invention used for the different technological operations (turning, planing, etc.) processes and has the possibility to change and use the different types of cutting tool. This device guarantees the changing of side rake and other angles of tool in process of cutting or before of this operation. The changing of angles in the tool makes possibility to decrease the wear of cutting tool and break the chip that improves the process of cutting in whole.

2. Assembling cutting tool of A.I.Rozenblat

UNION OF SOVIET SOCIALIST REPUBLICS

(19) **SU** (11) **1199466 A**
(51) 4 B 23 B 27/04

USSR STATE COMMITTEE ON INVENTIONS AND DISCOVERIES

SPECIFICATION OF AUTHOR'S CERTIFICATE

(21) 3766852/25-08
(22) 07.06.84
(46) 12.23.85. Bulletin №47
(72) A.I.Rozenblat
(53) 621.9.025 (088.8)
(56) US Patent №2627107, cl.29-96, published in 1950.

(54) ASSEMBLING CUTTING TOOL OF A.I.ROZENBLAT
(57)   An assembling cutting tool, for example a parting-off tool which comprises a holder in which is installed the cutting part with the main and auxiliary rear surfaces, and a support for interaction with the cut-off portion of a billet. The cutting tool is distinguished by the fact that in order to reduce its fretting, the support is made in the form of one or several solids of revolution and is located on one of the auxiliary rear surfaces of the cutting part of the tool.

The invention is related to metal processing. The aim of the invention is to increase the resistance of the assembling cutting tool due to reduced fretting of the cutting tool.

Figure 1 shows the suggested device, front view. Figure 2 shows the same, view from above; Figure 3 shows section A-A of Figure 1.

The assembling cutting tool contains the cutting part 1 on which are made the main cutting edge 2 and auxiliary cutting edges 3, and also their generating surfaces: front, rear 4 and auxiliary 5 and 6 which are turned to the fastened and the cut-off sides of the work piece respectively.

On the auxiliary rear surface 6, turned to the cut-off portion of an article, is made the slot 7 whose direction coincides with the diagonal 8 connecting the operating area of the lateral rear surface 6 with the vertex of the cutting edge 2 of the tool. Into the slot 7, for example, is inserted the rolling-contact bearing 9 in the manner of a cassette and is fastened in the separator 10. The rolling-contact bearing 9 can be fastened, for example with the aid of magnets 11 and 12(other ways of fastening are possible, for example with screws). On the cutting part 1 of the tool, on the main rear surface 4 of the edge are made tool angles 13 and 14, over the rear edges 5 and 6 are made tool angles 15 and 16.

The assembling cutting tool operates in the following manner:

The tool is installed in the tool holder and fastened there. The holder of the parting-off tool (or the main cutting edge 2) is installed strictly perpendicular to the axis of the article and in such a manner that the cutting edge 2 is installed, for example, along the axis of the article. Upon cutting, i.e., while performing the longitudinal feed to the tool, the auxiliary lateral cutting edge 3 is engaged into contact with the surface of the cut-off portion of the article. When the cutting edge 2 and 3 approaches the axis of the article, the surface of the cut-off portion under the effect of its own weight is pressed to the auxiliary rear surface 6 and, depending on diameter of a billet is engaged into contact with one of the rolling-contact bearings 9.

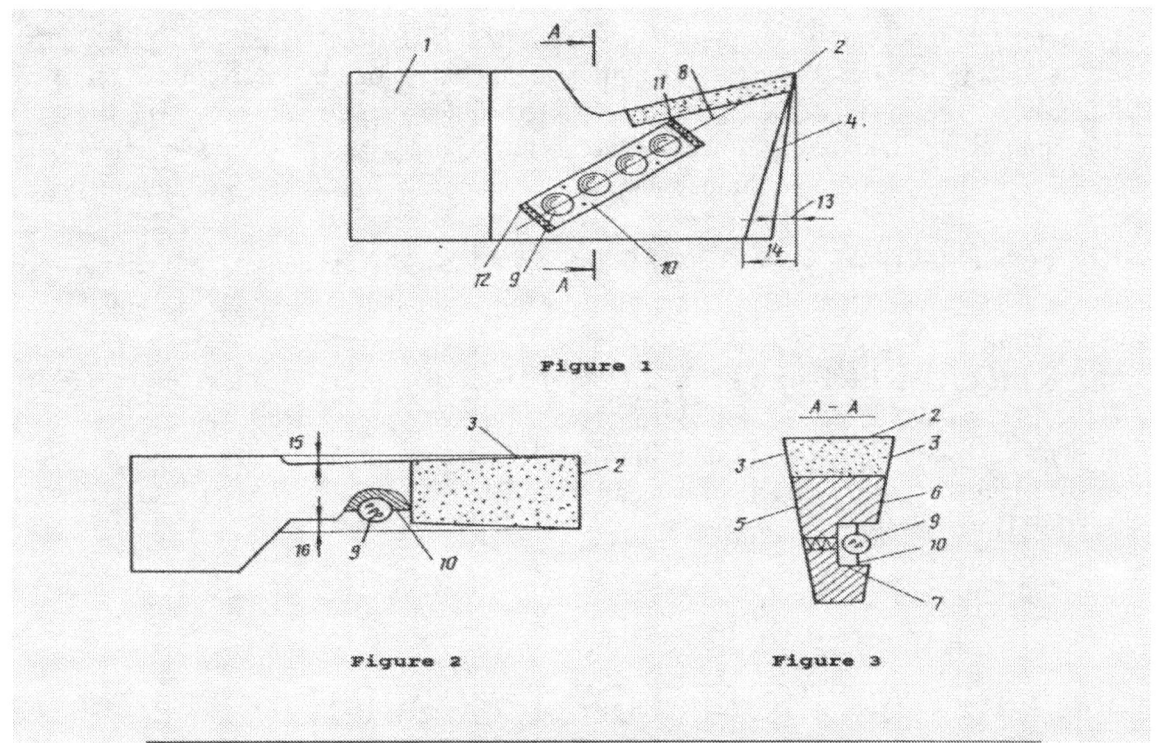

Figure 1

Figure 2   Figure 3

Assembling cutting tool of A.I.Rozenblat

The advantages and utility of this invention was evaluated by the Technology Targeting Incorporated (USA) it shown below and also, as example, some calculations are made by the author for the designing of parting-off tool.

EVALUATION OF THE INVENTION "REDUCED-FRETTING ASSEMBLING CUTTING TOOL" by Technology Targeting, Incorporated (USA).

NON-CONFIDENTIAL INVENTION SUMMARY
[Information Provided Without Obligation]

TTI #-291

REDUCED-FRETTING ASSEMBLING CUTTING TOOL

[Inventor: Anatoly Rozenblat]

INVENTION DESCRIPTION: Device to decrease the fretting which normally occurs in a metal cutting operation.

UTILITY AND ADVANTAGES: Fretting is a common problem in cutting and milling operations. Debris, usually in the form of oxides, accumulates in the narrow space between surfaces, causing gulling, seizing or fatigue cracks which grow as debris and stress increase. One of the recognized solutions to the fretting problem is to pressure forcing two surfaces together, but this has not been easily accomplished in a routine fashion in the metal cutting and fabricating industry.

This invention accomplishes an automatic reduction of fretting by providing alternative levels of support/pressure at the rear of a cutting tool. This is done by means of a holder with rolling-contact bearings. As the cutting edge is fed longitudinally and wears slightly, it is pressed into an auxiliary rear surface and come in contact with one of the rolling-contact bearings. This provides the additional support to reduce the fretting as the cutting surface wears.

PATENT AND DEVELOPMENT STATUS: This invention has been patented in the former Soviet Union; the inventor brought the technology and know—how with him when he immigrated to the USA.

FURTHER INFORMATION: Copies of the patent are available upon request. Data demonstrating its efficacy of use are available from the inventor under cover of a Confidential Disclosure Agreement. Commercial rights are available for exclusive, non-exclusive or field—of—use exclusive licensing. When making inquiries, PLEASE REFER TO THE TTI # SEEN ABOVE.

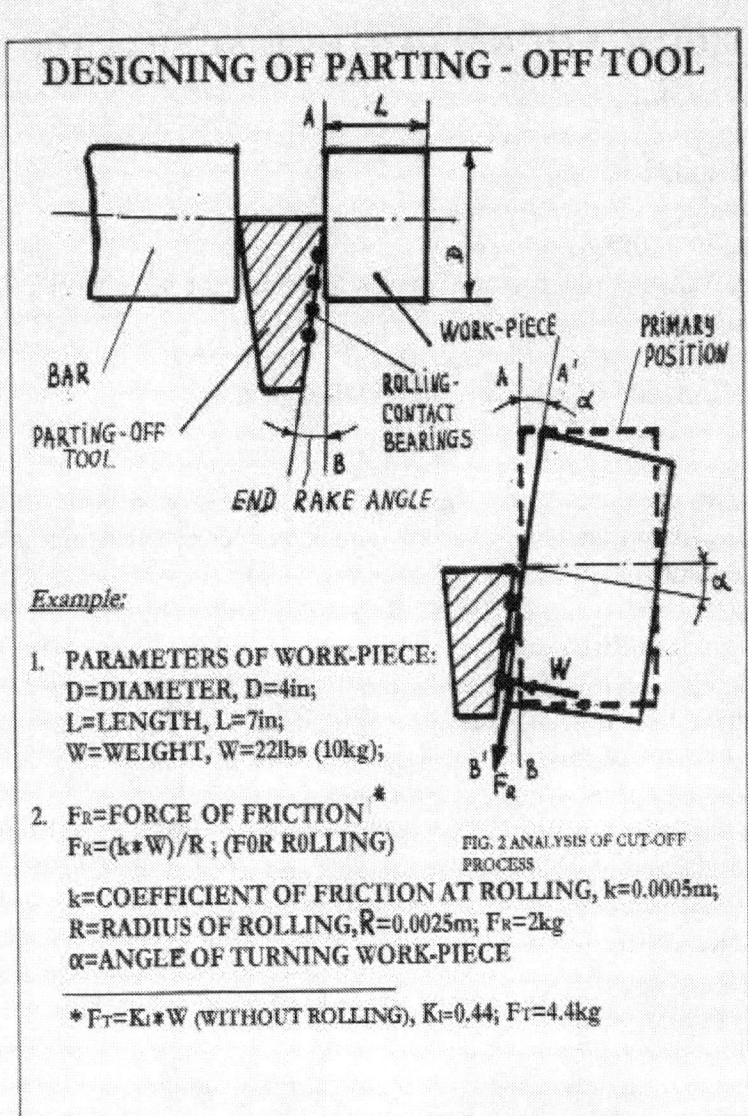

DESIGNING OF PARTING - OFF TOOL

BAR

PARTING-OFF TOOL

END RAKE ANGLE

WORK-PIECE

ROLLING-CONTACT BEARINGS

PRIMARY POSITION

Example:

1. PARAMETERS OF WORK-PIECE:
   D=DIAMETER, D=4in;
   L=LENGTH, L=7in;
   W=WEIGHT, W=22lbs (10kg);

2. F_R=FORCE OF FRICTION *
   $F_R=(k*W)/R$; (FOR ROLLING)

   FIG. 2 ANALYSIS OF CUT-OFF PROCESS

   k=COEFFICIENT OF FRICTION AT ROLLING, k=0.0005m;
   R=RADIUS OF ROLLING, R=0.0025m; F_R=2kg
   α=ANGLE OF TURNING WORK-PIECE

   ---
   * $F_T=K_1*W$ (WITHOUT ROLLING), K_1=0.44; F_T=4.4kg

3. Wrench of A.I.Rozenblat for jaw chuck

UNION OF SOVIET SOCIALIST REPUBLICS

(19) SU (11) 1505676 A1
(51) 4 B 23, B 31/00
USSR STATE COMMITTEE OF INVENTIONS AND DISCOVERIES

SPECIFICATION OF AUTHOR'S CERTIFICATE

(21) 4291302/25-08
(22) 28.07.87
(46) 07.09.89 Bull.#33
(75) A.I. Rozenblat
(53) 621.941.229.3 (088.8)
(56) Seminsiy V.K and other. Contrivances and tools for lathe works. Kiev, Technique, 1977, pp.22-23, fig.18.

(54) WRENCH OF A.I.ROZENBLAT FOR JAW CHUCK

ABSTRACT

(57) The invention relates to the machine—tool industry and more concretely to the device for handle clamp-unclamp of jaw chuck. Objective of this invention there is increasing of ergonomic qualities of installation.

The rod consists from two parts 1 and 2 it are joined on hinge by means of strip. The handle also consists from two parts 3 and 4 in it are fixed the sources 5 and 6 of power so that they connected by electric circuit with signal system 18 and interrupter 12 of circuit. On the rod of interrupter 12 actions the screw 19 so that when wrench rotates and this time said rod receives rotation and displacement from pulley 20, belt (pull rope) 21 and driving pulley 22 contacting with body of jaw chuck at the rotation of wrench.

In primary position the contacts of interrupter 12 are opened and when the effort of clamp is achieved, the screw 19 closes the contacts and in this time appear a signal (sound or light) showing about that clamp of jaw chuck is finished.

1 CLAIM; 2 DRAWING FIGURES.

DESCRIPTION

The invention relates to the machine-tool industry and more concretely to the device for handle clamp-unclamp of jaw chuck. Objective of this invention there is increasing of ergonomic qualities of installation.

On Fig.1 shown a wrench, axial section; on Fig.2-view A on Fig.1.

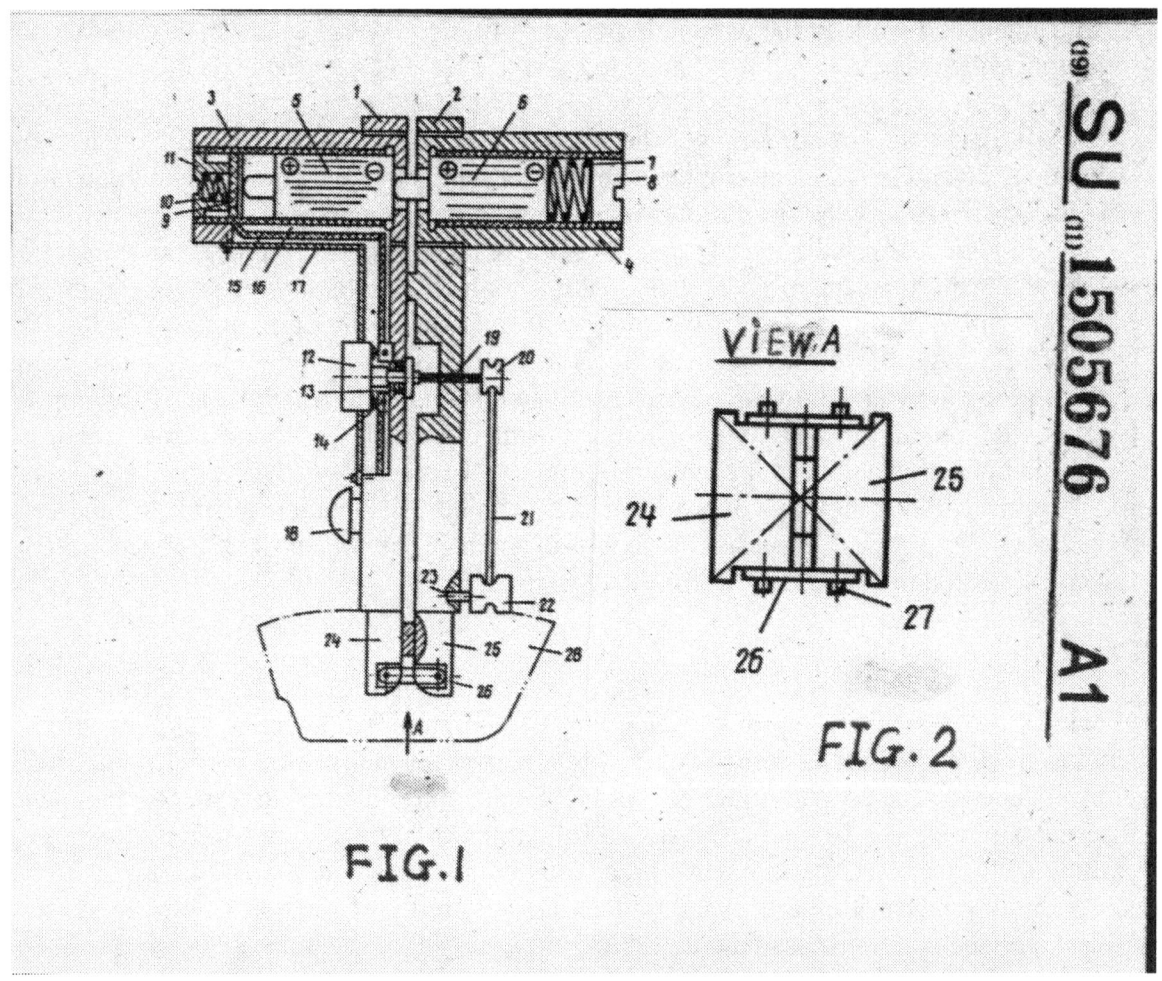

FIG.2

VIEW A

FIG.1

Wrench of A.I. Rozenblat for jaw chuck

Improvement of ergonomic characteristics in the design of new tools for machine manufacturing is the main objective of this device shown in Figure 1 and 2. The first steps to address the problem of designing safety tools for the blind or deaf-mute machine—tool operators.

As example of this design is a wrench of A.I.Rozenblat for a jaw chuck (patent #1505676) which consists of the left 1 and right 2 parts of a rod, to which are accordingly fixed the left 3 and right 4 parts of the handle. The sources 5 and 6 of power (a battery) are installed in these parts and generate contact with the aid of a spring 7, a plug 8 in the right part of the handle 4 and the conducted washer 9, as well as a spring 10 and plug 11 for the left part of handle 3.

In the rod 1 the circuit interrupter 12 is installed with a return spring 13 and conduction current contacts 14. Also, in the left part of the handle 3 and the rod 1electric wire 15 is situated in the channel 16 of the left part 3 and is closed in by the housing 17. On the left part 1 of the rod a sound or signal installation 18 is arranged and on the right of part a mechanism of driving is arranged consisting of a movable screw 19 fixed on the axis of the pulley 20 with the belt (a pull rope 21) and a driving pulley 22 fixed on the axis 23.

Parts 1 and 2 accordingly have the different shanks 24 and 25which are fixed by hinge 26 to the axis 27.

- Adjustment and work by the wrench in period of fixing of detail in jaw chuck realizes by the following way:

   1. In right part 2 of rod arranges the screw 19 and then both parts 1 and 2 connect on the hinge 26 and axis 27. Interrupter 12 by screw stands in open condition and power on signal installation 18 absences and wrench is ready for the operation.

   2. In period of installation of wrench in square of jaw chuck 28 provisions contact of pulley 22 with the plane of jaw chuck 28 and later after of twisting the pulley 22 gives a rotation through the belt 21 to the pulley 20. And in result of rotating of pulley 20 transfers a progressive movement to the screw 19 (if the screw 19 moves to the right side) which the rod 1 moves and closes contact 14 and after of finishing of clamp of detail (motion of screw 19 controls in adjustment)provides a sound signal about of finishing of clamp and need it excavation from the square of jaw chuck.

   So, the safety conditions of work for the machine-tool operator guarantees and also decreases the physical expenses on fixing of detail in jaw chuck and this accordingly increases ergonomic qualities of installation.

CLAIM:

Wrench for jaw chuck arranged on it body and comprising rod and handle *distinguishing* that with the objective of improving condition of operation said wrench supplies unified electrical electric circuit with sources of power, and signal installation, interrupter of circuit and mechanism action on interrupter so that the handle and rod is made from two parts and these said sources of power and are installed in both parts of handle, and interrupter, signal installation which are fixed at one part of rod, and the mechanism action on interrupter at the other part and joined with the first part so that has possibility of rotation and made in view of spiral pair and belt transmission with driving pulley which is used for connection with the body of jaw chuck in the process of clump-unclump of detail.

NON-CONFIDENTIAL INVENTION SUMMARY

AIR # 297

WRENCH OF A.I.ROZENBLAT FOR JAW CHUCK

(Inventor: Anatoly Rozenblat)

INVENTION DESCRIPTION:

Device is destined to increase the ergonomic characteristics of installation and also decrease the expenses on fixing of detail and finally to guarantee the safety conditions of work for the machine-tool operator.

UTILITY AND ADVANTAGES:

The present wrench in generally for fixing of bar in jaw chuck does not guarantee safety of machining and particularly this concern for the blind or deaf-mute machine-tool operators.

This inventions removes these defects so that wrench supplies unified elements of electrical circuit with the sources of power, and signal installation, interrupter of circuit and mechanism action on interrupter so that the handle and rod is made from two parts and these said sources of power are installed in both parts of handle, and interrupter and signal installation are fixed at one part of rod, and mechanism action on the interrupter is installed at other part and joined with the first part so that has possibility of rotation and made in view of spiral pair and belt transmission with driving pulley used for connection with body of jaw chuck in the process of clump-unclamp of detail.

PATENT AND DEVELOPMENT STATUS:

This invention has been patented in the Soviet Union; the inventor has brought the technology and know how with him when he immigrated to the USA.

FURTHER INFORMATION:

Copies of the patent are available upon request. Data demonstrating its efficacy of use are available from the inventor under cover of a Confidential Disclosure Agreement. Commercial rights are available for exclusive, non-exclusive or field-of-use exclusive Licensing.

When making inquiries, PLEASE REFER TO THE AIR # SEEN ABOVE.

The advantages of Rozenblat's wrench for jaw chuck are shown on the separate page.

ADVANTAGES OF ROZENBLAT'S WRENCH*

1.   UNIVERSAL (USING FOR DIFFERENT TYPES OF JAW CHUCKS WITH HANDLE DRIVE).
2.   PORTABLE (ALL ELEMENTS ARE FOLDING)

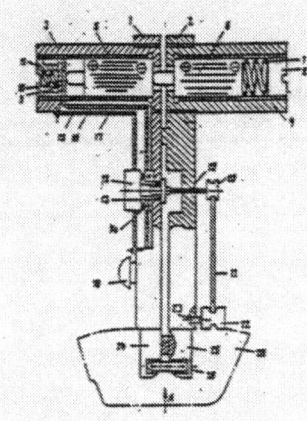 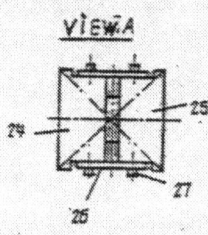

FIG. 3 ROZENBLAT'S WRENCH FOR JAW CHUCKS

3.   SAFETY FOR MACHINISTS WHO HAVE THE PHYSICAIL DEFLECTIONS.
4.   UNIFIED (PARTS AND UNITS ARE INTERCHANGEBILITY).

----

* COMPARED WITH CHUCK KEY SERIES HK40 (CATALOG №85)

4. A tool cleaning device

UNION OF SOVIET SOCIALIST REPUBLICS

(19) **SU** (11) **1131634 A**
3 (51) b 23 Q 11/02

USSR STATE COMMITTEE ON INVENTIONS AND DISCOVERIES

SPECIFICATION OF AUTHOR'S CERTIFICATE

(21) 3435998/25-08
(22) 05.10.82
(46) 12.30.84 Bulletin №48
(72) A.I.Rozenblat
(53) 621.93.02 (088.8)
(56) 1.USSR Author's Certificate
№465310, Cl.B23 Q 11/02,1972 (prototype)

(54)   (57) **A TOOL CLEANING DEVICE** which comprises a chip cleaning device made in the form of a spring actuated bar installed with the possibility of reciprocating motion. The device is distinguished by the fact that in order to increase efficiency of cleaning the tool, it is equipped with second bar installed with the possibility of reciprocating motion oppositely to the first, made identically and connected with the first bar. In addition, the bars, having with channels made additionally in them and designed for supplying compressed air to the cleaning zone, are installed over faces of the tool which also have additionally made channels, with an entry to the face surfaces of the tool and with an exit into the chip groove of the tool.

The invention is related to mechanical processing, namely to cutting billets of predominantly cold rolled products, particularly in operation on milling cutting-off machines.

A tool cleaning device is known which comprises a chip cleaning device made in the form of a spring actuated bar installed with the possibility of reciprocating motion (1).

The fault of the known device consists in impossibility of using the said design for cleaning chips from chip grooves of a tool at mechanical cutting of cold rolled products with circular saws.

The goal of the invention is to increase efficiency of cleaning tools. The formulated goal is achieved by the fact that the tool cleaning device, which comprises a chip cleaning device made in the form of a spring actuated bar installed with the possibility of reciprocating motion, is equipped with second bar installed with the possibility of reciprocating motion oppositely to the first one, made identically to the first bar and connected to the latter. In addition, the bars, with additionally have made channels designed for supplying compressed air to the cleaning zone, are installed over the faces of the tool which also have additionally made channels, with entrance to the face surfaces of the tool and with exit to the chip groove of the tool.

The drawing shows the diagram of the suggested device for cleaning tools, using a segmental saw as an example.

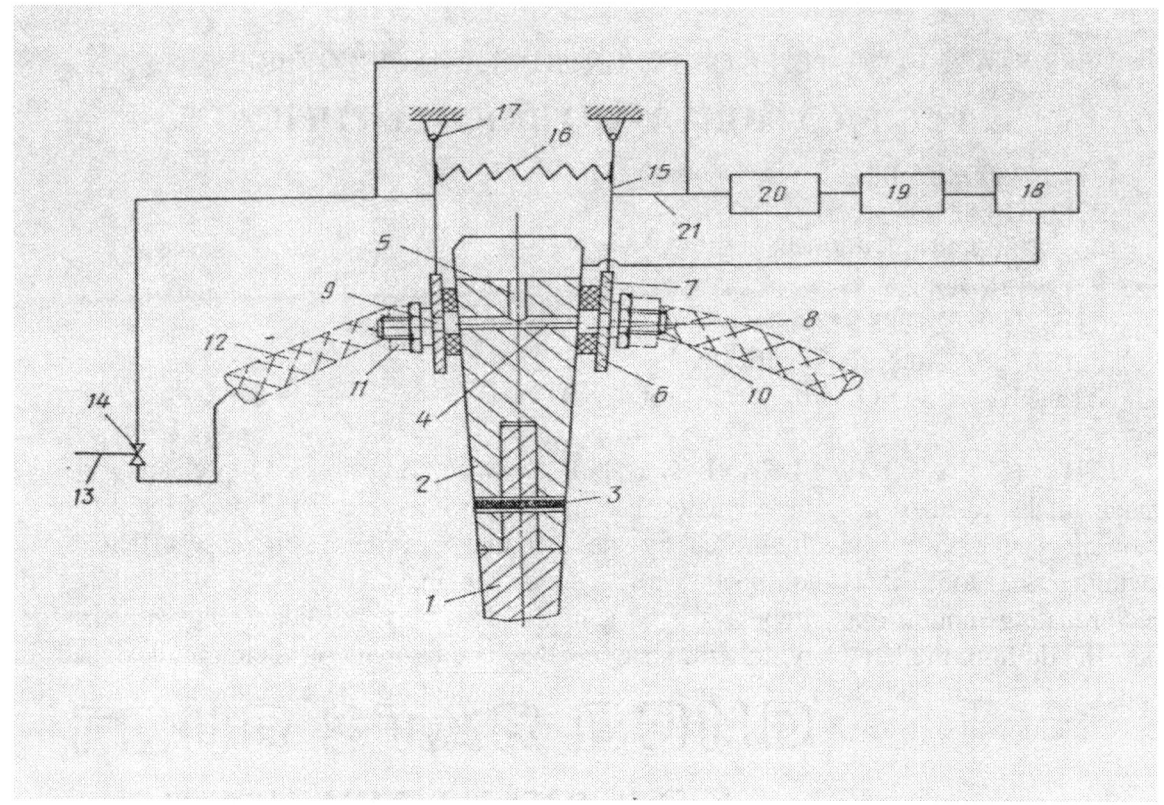

Tool cleaning device

The circular saw 1 is assembled with segments 2 with the aid of rivets 3. In the segment are drilled a transverse 4 and longitudinal 5 channels, and from lateral surfaces of a tooth of the saw segment are brought cleaning bars 6, for example from a soft material in the form of a brake band which are fastened on the metallic lining 7. The body of the cleaning bars 6 and the lining 7 contains channels 8 and 9. A union 10 is screwed into the body of the metallic lining 7. Then the coupling nut 11 is fastened, as well as the flexible hose 12 through which is supplied compressed air to the cleaning zone from the system 13 through the valve 14. The cleaning bars 6 are fastened with the aid of levers 15 interconnected by the spring 16 and the jointed tie 17. A pickup 18 is installed whose action is transmitted to the actuating element 20 and the control element 21.

Operation in the mode, i.e when chips are removed directly from the cutting zone, that is from the front surface of the tooth recess of the segmental saw is conducted in the following manner.

The circular segmental saw 1 is installed on the milling cutting-off machine, the cleaning bars 6 are installed in free position in relation to the lateral surface of the saw, i.e in arbitrary manner. Then, with the saw installed fixedly, the cleaning bars 6 are arranged and adjusted by means of the joints 17 and the levers 15 in such a way that the planes contact and match on of the teeth. Here, the transverse channel 4 of the tooth and the channel 8 of the bar 6 are matched and the bars 6 are fixed in this position with the aid of the levers 15 in the jointed tie 17.

The pickup 18 (for example, a time relay) is set for a period equal, for example, to matching of the next channel 4 of the segment tooth of the circular saw. The matching cycle of channels 4 and 8 and adjusting the pickup 18 are determined experimentally. Then, the circular saw 1 is set into rotary motion.

At the moment of matching the channel 4 of the next tooth of the saw segment and the channel 8 of the cleaning bar 6, the pickup 18 of the time relay actuates, i.e during this period the cleaning bars 6 are pressed to the lateral surfaces of the circular saw due to the fact that action of the pickup 18 is transmitted to the amplifier 19, actuating element 20 and control element 21 (made in the form of a lever or electromagnet or other mechanism). At the same time, the electromagnetic valve 14 actuates, and air from the system 13 through channels 8,9,4 and 5 comes to the cutting zone, i.e directly under the formed chips over the front surface and the recess of the segment tooth of the saw. Here, com-pressed air is supplied by pulse in an automatic mode and the spring 16 is in compressed state.

Utilization of the suggested device increases resistance of cutting tools and efficiency of cutting-off operations with circular segmental saws, particularly in cutting of viscous, stainless and other materials.

TECHNOLOGY TARGETING, INCORPORATED

A Non-Profit Scientific/Educational Organization

2940 South Warr
Salt Lake City, UT 84109
(801) 487-9800

NON-CONFIDENTIAL INVENTION SUMMARY
[Information Provided Without Obligation]

TTI# - 289
TOOL CLEANING DEVICE
[Inventor: Anatoly Rozenblat]

INVENTION DESCRIPTION: Automatic removal of chips or fragments (e.g., of metal from a cutting or milling operation) from a circular cutting tool.

UTILITY AND ADVANTAGES: Fragments produced during cutting or milling of metal can collect and damage either the cutting edge of the tool being used or the surface of the material being processed. This device uses two spring-actuated bars in opposing reciprocal motion, placed over the faces of a tool. The tool itself is given specific "cleaning grooves" along transverse and longitudinal axes; the cleaning device then sweeps chips and fragments into these cleaning grooves and out from the tool. The actuation of the pickup motion is set to match the revolution of the tool, so that it is always synchronized with the appearance of the cleaning grooves.

The invention increases resistance and efficiency of cutting tools, such as circular segmental saws, particularly in cutting of viscous, stainless, and related materials.

PATENT AND DEVELOPMENT STATUS: This invention has been patented in the Soviet Union; the inventor has brought the technology and know-how when he emigrated to the US.

FURTHER INFORMATION: Copies of the patent are available upon request. Data demonstrating its efficacy of use are available from the inventor under cover of a Confidential Disclosure Agreement. Commercial rights are available for exclusive, non-exclusive or field-of-use exclusive licensing. When making inquiries, PLEASE REFER TO THE TTI # SEEN ABOVE.

5. Installation of A.I.Rozenblat for cutting of metal bar

UNION OF SOVIET SOCIALIST REPUBLICS

(19) SU (11) 1504064 A1
(51) 4B 23Q 41/04; B21 J 7/00, B 21 D 28/00
USSR STATE COMMITTEE OF INVENTIONS AND DISCOVERIES

SPECIFICATION TO AUTHOR'S CERTIFICATE

(21) 4305946/25-27
(22) 14.09.87
(46) 30.08.89 Bull.#32
(75) A.I.Rozenblat
(53) 621.96 (088.8)
(56) Author's certificate of the USSR #683847 cl. B21 J 11/00,25.07.77
(54) INSTALLATION OF A.I.ROZENBLAT FOR CUTTING OF METAL BAR

ABSTRACT

(57) Invention applies to the metal machine equipment for cutting of metal billet particularly to the press-forging plant for the piercing-punching of sheet metal blanks.

Objective of this invention is the decreasing of industrial power in manufacturing processes. As example, the process includes the cutting—off machine and stamping press for piercing or punching.

Cutting machine 1 is placed on the upper deck and stamping press 2 on the lower deck. Between of these machines 1 and 2 there is a guide 3 and piston 4. There is a window at the wall of this guide 3 it is closed by the mobile shut-off damper 16. Upper surface of the piston 4 has displacement 9.

On the cutting-off machine 1 from the bar 6 cut-off some part 5 it drops later into guide 3 on skewed front surface of piston 4. This piston is jointed with the rod 7 and impacted element 8 so that acts on the upper part 20 of stamping press 2 which makes the useful work.

So, the cut-off part 5 from the billet 6 passing through the guide 3 and connection with the piston 4 later falls down to the roller conveyer 26 and further to the container 27.

5CLAIMS;3 DRAWINGS FIGURES

Description

The objective of this invention is the decreasing of power consumption in manufacturing processes. This factor is more important for the cut-off processes of large diameter stainless and high-resistance steels while using cold circular segmental saws. The application of the new combined installation of A.I.Rozenblat (patent #1504064 in Russia) is shown in general view on Figure 1, Figure 2 in view of arrow A on Figure 1 and also on Figure 3 in view of arrow B on Figure 1.

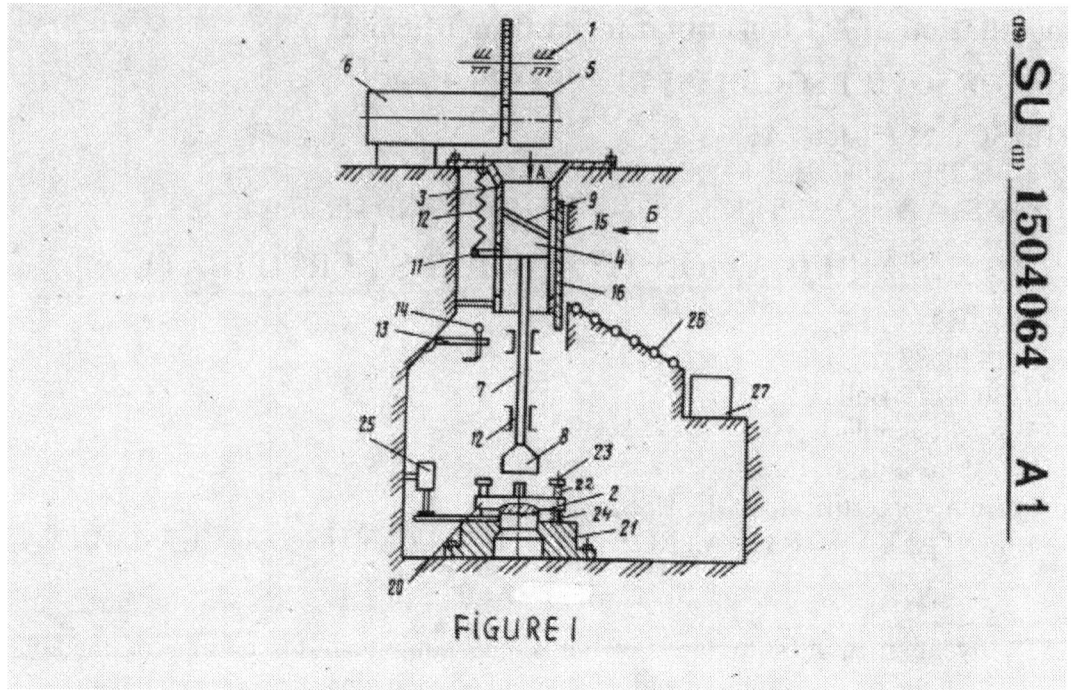

FIGURE 1

SU (19) (11) 1504064 A1

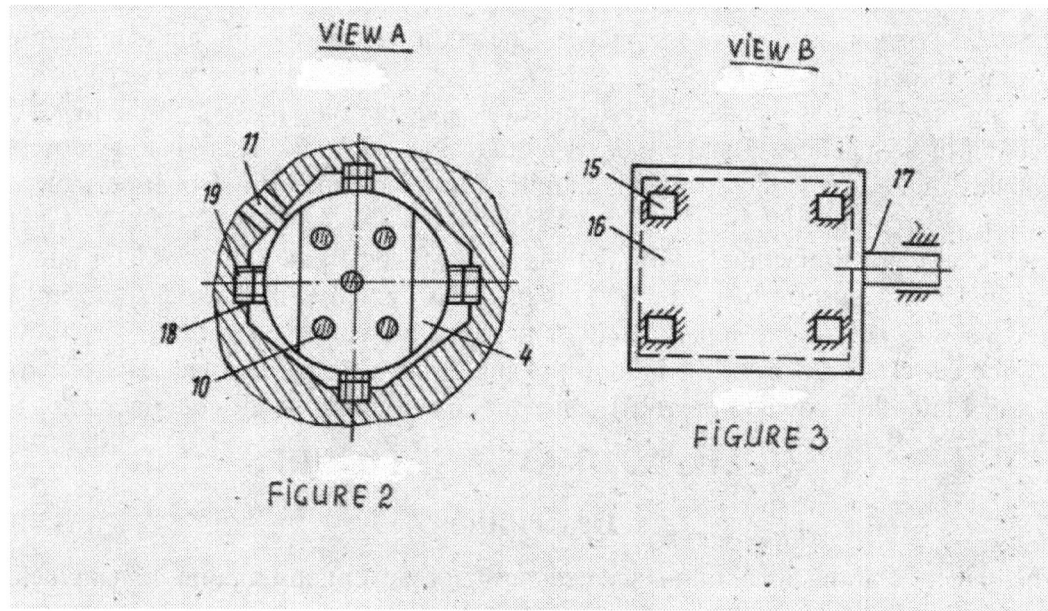

VIEW A

FIGURE 2

VIEW B

FIGURE 3

Installation of A.I.Rozenblat for cutting of metal bar

This invention permits to use the kinetic energy of the free-falling cutting part of the bar for conversion into useful energy, for instance for the moving of a die-forging press or other machine.

The installation for the cut-off process of metal bars contains a machine 1 which is arranged on the upper deck. Placed on the lower deck, in a trench under the industry building, is the other cutting machine-such as a stamping press 2 for the piercing or punching of the detail from sheet metal. Between the upper and lower decks, the guide 3 for the transporting of the piston 4 is mounted.

The guide 3 is placed under the zone of arrangement of the cutting part 5 of the billet 6. The end of the guide 3 locked to the cutting part 5 has the form of a funnel and is lined with rubber. The piston 4 is joined with the rod 7, on which is fixed the striker element 8. The upper face of the piston 4 has displacement 9. On his face, the damping elements and rolling-contact bearing 10 is set up.

On the side of the piston 4 is fixed the stop 11which is joined with the spring 12 and the flange of the guide 3. The micro-switch 14, which regulates the movements of the piston 4, is situated on the arm 13 under the stop 11.

In the wall of the guide 3, on the side of the lower end displacement 9 of the piston 4 in the side guards 15 the shut-off damper 16 is placed. The shut-off damper is supplied by the drive 17which is not shown on the Figure 3.

The guide 3 is secured with longitudinal grooves 18 that support the direction of the piston 4. So, the piston 4 is in view of some parts which could be regulated with the elements 19 that enter the grooves 18. As a result of this, the sizes of the cross-section of the piston 4 could be changed.

The stamping press 2 installed on the lower deck has a movable upper part-the punch 20—and an immovable lower part—the die 21-as well as the guide columns 22 with the shoulder 23 joining the two parts 20 and 21;the spring 24 that return the punch to its in normal upper position; and finally the facility 25 for servicing the sheet metal (not shown) in the working zone of the stamping press 2.

For the removal of the sheet metal, the installation is equipped by the roller conveyor 26 and container 27.

- The installation works the following way:

On the machine 1 cut-off part 5 from the bar 6 and then this part 5 falls into the guide 3. The piston 4 before of falling part 5 into the guide 3 is fixed by the spring 12. In during of falling part 5 the last moves piston 4 in account of shock. The strength of this shock is suppressed by the liquid dampers. The piston 4, rod 7 and shocked elements 8 are lowering and action on the upper part 20 of stamping press 2 and lower its to the guide of columns 22.

At the interaction of parts 20 and 21 of stamping press 2 make a cutting process, for example the punching of detail from the sheet metal which is placed on the part 21. After of finishing of punching process, switches the time-relay (not shown) and moves the shut-off damper 16, opens window in the wall of guide 3 so that part 5 moves on the skewed end of the piston 4 and further to the roller conveyer 26 and to the container 27.

The spring 12 raises piston 4 in normal primary upper position and spring 24 raise in normal position a movable part 20 of stamping press 2. And further the cycle of this process repeats again.

SAMPLE:

*In manufacturing process is used the cutting-off machine;
*The bar is made in view of round billet of diameter 200 millimeter from carbon steel;
*The sheet blank is made from the steel sheet by thickness =0.50 mm;
*The depth of dropping the cut of bar is equal 6.0 meter;
*The length of cut part of bar is equal 500.00 millimeter.

Solution:

Using of this invention as example at given above data allowed to punch for one shock of die about of 80 pieces diameter 5.0 mm.

CLAIM:

1. Installation for cutting of metal bar, comprising two placed one under another cutting machines, piston and guide for vertical displacement which is arranged between these said machines so that bottom machine is fulfilled from two parts-lower and upper fixed with possibility of interaction with the piston and displacement to the direction so that its coincides with direction of displacement piston, distinguishing that with the objective of reducing of industrial power system, in the capacity of cutting machine situated on the upper deck is used the machine for cutting of part bar so that said guide is placed under zone of displacement of cutting part and piston is fixed with possibility axial of displacement relatively said cutting machine standing on the upper deck so that in the wall of this guide is made the window for removal of cut-off bar and upper end face of piston is made with displacement which is turned by lower end to the window, and on this said end face are fixed the liquid dampers.

2. Installation on claim 1 distinguishing that on the upper surfaces of liquid dampers are set up the supports of rolling.

3. Installation on claim 1 distinguishing that from outer side of guide in zone of displacement window is arranged the shut-off damper with possibility of movement.

4. Installation on claim 1 distinguishing that piston supplied with means of limited part on its displacement along guide which are fulfilled in view of spring connecting said piston with the guide and micro-switch.

5. Installation on claim 1 distinguishing that piston is made by complex from the parts having a vertical plane of joint which are installed with possibility of controlling displacement to the direction perpendicularly of axis piston.

The advantages and utility of the invention "Installation of A.I.Rozenblat for cutting of metal bar" shown on the next pages.

NON-CONFIDENTIAL INVENTION

AIR # 295

INSTALLATION OF A.I.ROZENBLAT FOR CUTTING OF METAL BAR

(Inventor: Anatoly Rozenblat)

Invention description:

Objective of this device is the decreasing of power in manufacturing processes particularly for the cut-off machines.

Utility and advantages:

Milling and cut-off machines with using of segmental saws recommends for this invention advantageously in manufacturing processes.

This invention utilizes the kinetic energy free falling cut part from the big bar and makes the useful work. And besides this installation comprises a milling-cutting off machines for cutting of bar which is placed on the upper deck in manufacturing process and other machines, for example a die for piercing or punching is fixed on the lower deck.

After of finishing of cut-off the heavy part the last under of action of gravity drops down in special installation and makes the useful work in any machine it installed in lower deck.

Patent and development status:

This invention has been patented in the Soviet Union and the Inventor brought the technology and know how with him when he immigrated to the USA.

Further information:

Copies of the patent are available upon request. Data demonstrating its efficacy of use are available from the inventor under cover of a Confidential Disclosure Agreement. Commercial rights are available for exclusive, non -exclusive or field-of use exclusive licensing.

When making inquiries, PLEASE REFER TO THE AIR # SEEN ABOVE.

ADVANTAGES OF INSTALLATION

1. **DECREASING OF POWER CONSUMPTION.**

2. **SIMPLICITY AND UNIVERSAL.**

3. **AUTOMATIZATION.**

4. **DECREASING OF COST AND LABOR.**

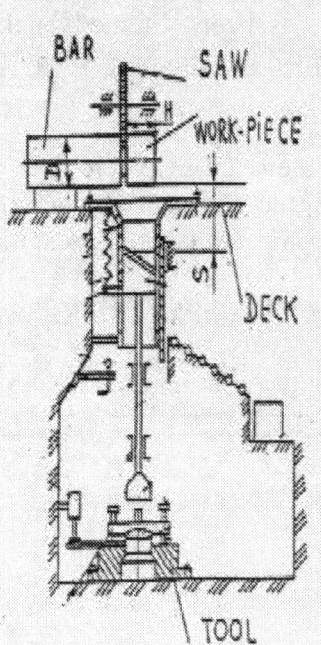

FIG. 1 SCHEME OF PROCESS

Example:

1. WEIGHT OF WORK-PIECE $W = V_1 \cdot \gamma$
   WHERE

   V_1 = VOLUME OF BAR, cm³ ; $V_1 = 0.75 D^2 H$

   γ = SPECIFIC WEIGHT OF MATERIAL, g/cm³
   H = HEIGHT OF BAR, cm
   D = DIAMETER OF BAR, cm

   D = 30cm; H = 5cm; γ = 7.859 g/cm³; W = 27.73kg (61 lbs.)

2. FORCE OF BLOW (USEFUL WORK) FROM
   FALLING BAR $F_B = W * S$, WHERE
   S = HEIGHT OF INSTALLATION, M   (S = 15M)
   SO, $F_B = 27.73*15 = 416$ NEWTONS = 0.42KN (3002.4ft-lbs)

6. Pipeline bend

UNION OF SOVIET SOCIALIST REPUBLICS

(19) SU (11) 1386786 A1
(51) 4 F 16 L 58/16, B 65 G 51/18

USSR STATE COMMITTEE OF INVENTIONS AND DISCOVERIES

SPECIFICATION OF AUTHOR'S CERTIFICATE

(21) 4125878/25-08
(22) 10.10.86
(46) 04.07.88 Bulletin №13
(71) Scientific Production Association Spetstekhosnastka
(72) A.I.Rozenblat, D.B.Kopanev and A.I.Vasnev
(53) 621.421 (088.8)
(56) USSR Author's Certificate №304356, cl. F 16 L 55/24,1970

(54) PIPELINE BEND

(57) The invention is related to pipeline technology for pneumatic transport of loose ferromagnetic materials. The aim of invention is to increase wear resistance of the bend walls. For this purpose, the body of the bend is made from nonmagnetic material, rolling-contact bearings are fastened diametrically on the body, a yoke with magnets is installed on the bearings. The yoke is set into rotation around the bend generating a rotating magnetic field inside the latter.

The ferromagnetic material moves inside the bend over the spiral-shaped trajectory revolving uniformly around the internal walls.

The invention is related to the pipeline technology for pneumatic transport of loose ferromagnetic materials.

The aim of the invention is to increase wear resistance due to increase wear resistance due eliminating wear of the bend walls.

Figure 1 shows the bend, transverse section; Figure 2 shows section A-A of Figure 1.

23

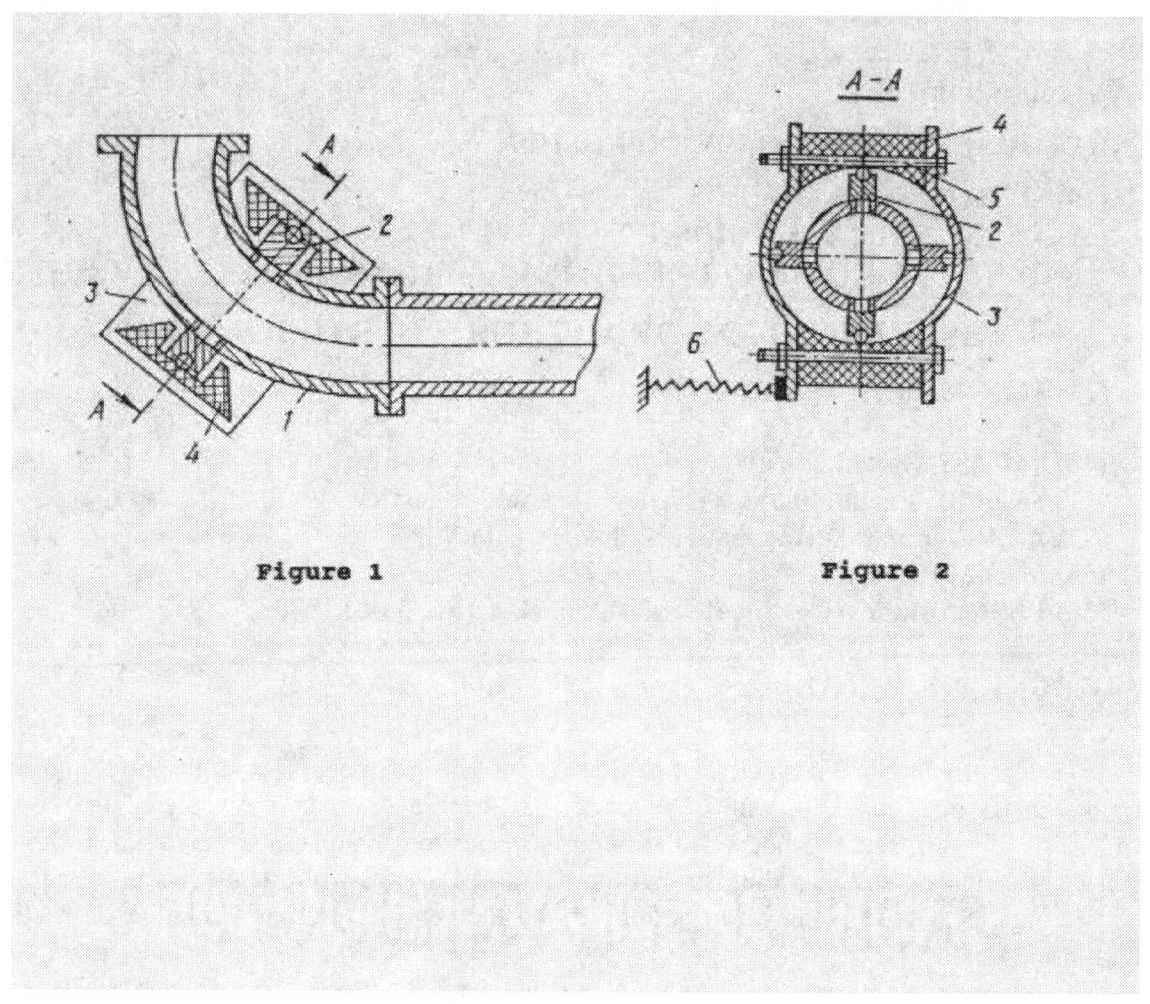

Figure 1   Figure 2

PIPELINE BEND

As shown on Figures 1 and 2 this device utilizes a non-metallic segment of the pipeline combined with a magnetic field created by the contact of bearings and yokes. Once the yoke is set in motion, it creates a rotating magnetic field which moves the ferromagnetic material in a uniform pattern, thus reducing contact with the pipeline surface and the resulting abrasion.

The pipeline bend includes such elements as the body of the bend 1 and the height-adjustable rolling- control bearings 2which are fastened diametrically on the body 1 of the bend. The body 1 is made from non-magnetic material. On the rolling-contact bearings 2 a roller cage 3 is installed in which magnets 4 are fastened by means of the fastening device 5. The balanced system is set into rotation with the aid of the impact pulse mechanism 6.

During pneumatic transportation of loose ferromagnetic material, under the effect of the magnetic field of the magnets 4, material moves inside a bend over the spiral-shaped trajectory revolving around the internal bend walls which reduces substantially the bend walls wear.

TECHNOLOGY TARGETING, INCORPORATED

A Non-Profit Scientific/Educational Organization

2940 South Warr
Salt Lake City, UT 84109
(801) 487-9800

NON-CONFIDENTIAL INVENTION SUMMARY
[Information Provided Without Obligation]

TTI# - 290
WEAR-RESISTANT PIPELINE ELBOW
[Inventor: Anatoly Rozenblat]

INVENTION DESCRIPTION: Device to decrease abrasion of elbows and turns in pipelines transporting ferromagnetic material.

UTILITY AND ADVANTAGES: Pipelines are used for pneumatic transport of a variety of materials, including metallic particles such as iron or steel – either as part of an industrial process or as a method of waste accumulation and processing. Such metal particals have a very high abrasion index, causing continual wear and damage to the pipeline which transports it. This damage is most pronounced at the elbows and curves of the pipeline, when the differential inertia of the particles ends them against the pipeline wall.

This invention utilizes a non-metallic segment of the pipeline, combined with a magnetic field created by contact bearings and yokes. Once the yoke is set in motion, it creates a rotating magnetic field which moves the ferromagnetic material in a uniform pattern, thus reducing contact with the pipeline surface and the resulting abrasion.

PATENT AND DEVELOPMENT STATUS: This invention has been patented in the Soviet Union; the inventor brought the technology and know-how with him when he emigrated to the US.

FURTHER INFORMATION: Copies of the patent are available upon request. Data demonstrating its efficacy of use are available from the inventor under cover of a Confidential Disclosure Agreement. Commercial rights are available for exclusive, non-exclusive or field-of-use exclusive licensing. When making inquiries, PLEASE REFER TO THE TTI # SEEN ABOVE.

Advantages of this invention "Pipeline bend" are shown on this page below

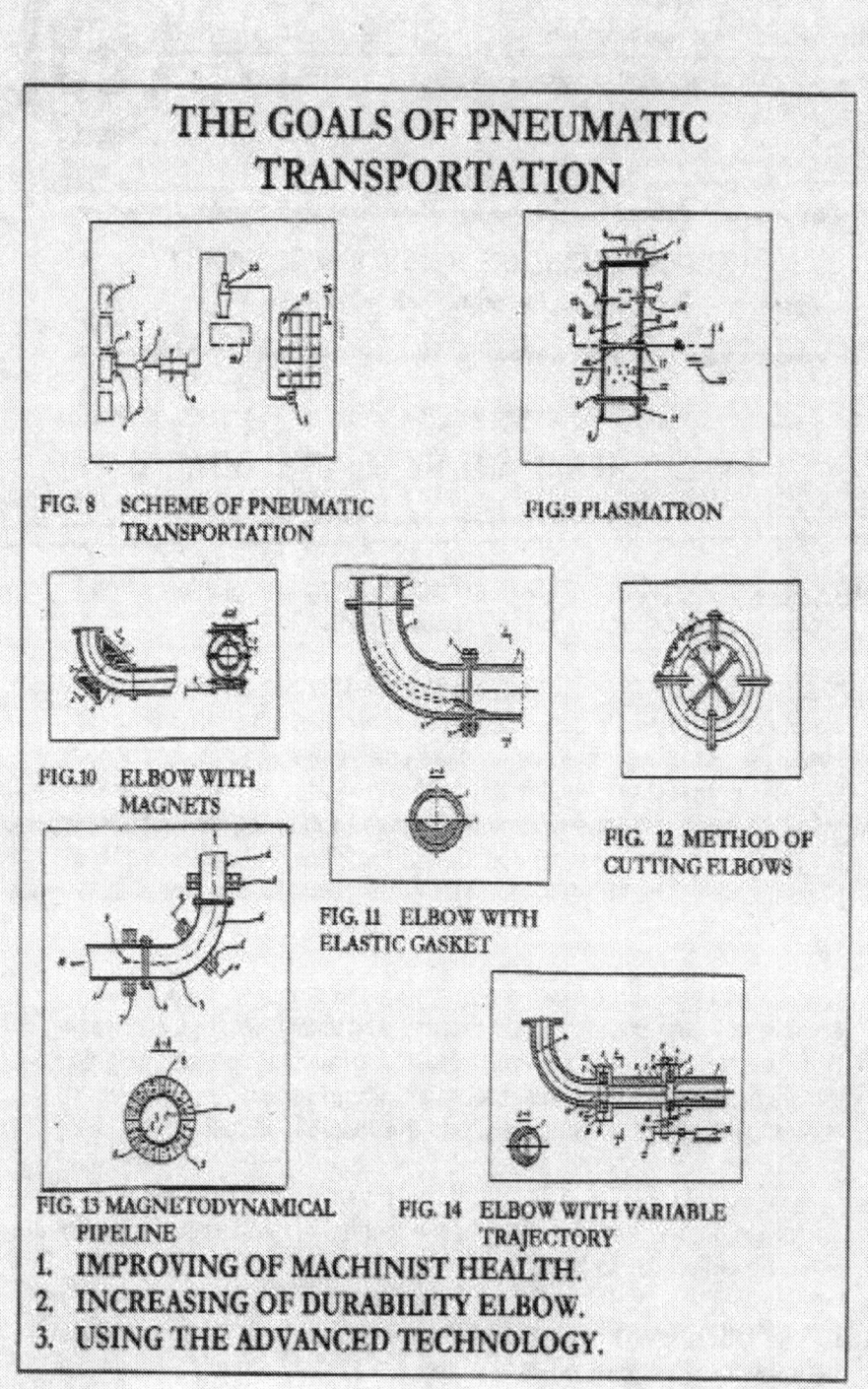

THE GOALS OF PNEUMATIC TRANSPORTATION

FIG. 8   SCHEME OF PNEUMATIC TRANSPORTATION

FIG.9 PLASMATRON

FIG.10   ELBOW WITH MAGNETS

FIG. 11   ELBOW WITH ELASTIC GASKET

FIG. 12 METHOD OF CUTTING ELBOWS

FIG. 13 MAGNETODYNAMICAL PIPELINE

FIG. 14 ELBOW WITH VARIABLE TRAJECTORY

1. IMPROVING OF MACHINIST HEALTH.
2. INCREASING OF DURABILITY ELBOW.
3. USING THE ADVANCED TECHNOLOGY.

7. Lathe tool

UNION OF SOVIET SOCIALIST REPUBLICS

(19) SU (11) 360155
(51) B23 b 29/04

USSR STATE COMMITTEE OF INVENTIONS AND DISCOVERIES

SPECIFICATION TO AUTHOR'S CERTIFICATE

(21) 1364370/25-8
(22) 15.08.69
(46) 1972 Bull.#36
(75) A.I.Rozenblat
(53) 621.9.025 (088.8)

(54) LATHE TOOL

(57) At present in the manufacturing industry, machining operations widely use right and left turning tools and some of their different modifications. At the same time, each type of the above-named tools corresponds to a certain size of shank. Analysis of such conditions shows that the specific consumption of material and manufacturing cost increase more than two times in comparison with the suggested new construction of turning and boring tools with the universal shank shown in drawings.

On Figure 1 is shown the lathe tool and in Figure 2—view on top.

The peculiarity of this cutting tool has the advantage that it is made separately, and it includes the shank 1 and head 2 and also the variable cut 3 with insert, which is fixed on the front end of the joint support 4 with aid of a "swallow tall" 5, and strips 6 which are fixed by a stud 7.

The joint support 4 is fixed on the conical axis 8. On one of end face of the joint support 4 are the teeth of a file 9 which engage with the teeth of a file of the ring 10 which is fixed by the nut 11.

The ring 10 is connected with the shank 1 of the tool by a screw 12. The turning of the joint axis around this axis accordingly changes the side—cutting edge angle. With the objective of objective of suppressing load from the machining process, the shank 1 is arranged in the spring damper 13.

CLAIM:

Lathe tool comprising a holder in which is fixed a head said tool distinguishing that with objective securing of possibility installation of head under different angles to the cutting surfaces, said head is fixed in holder of tool on conical axis which is tighten in conical hole of head and fixed in given position by means of screw.

In diagram is shown the changing of the cost job and material for the making of one set the prototype and this invention.

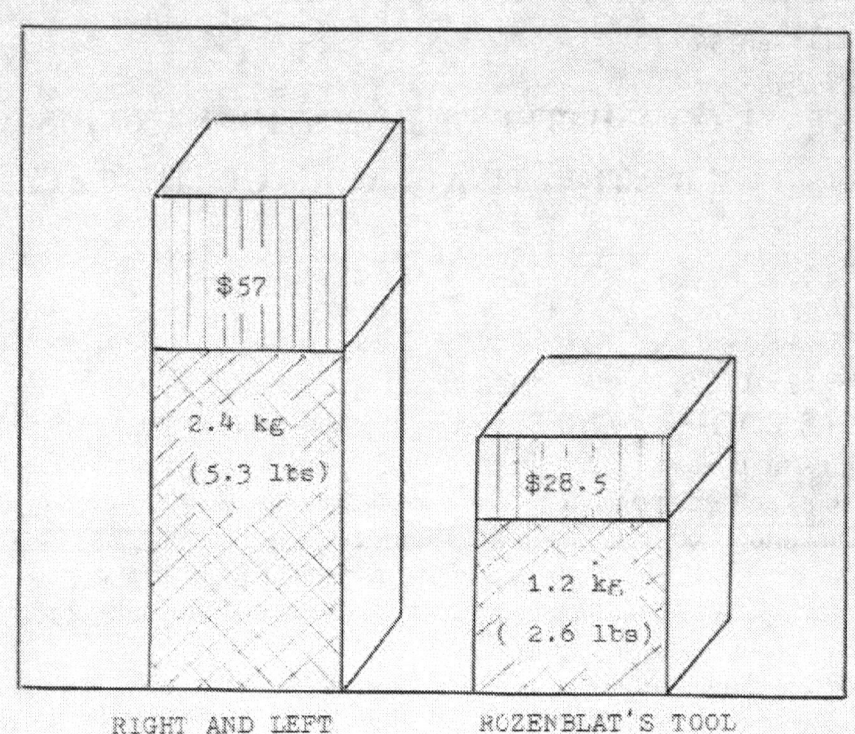

DIAGRAM OF CHANGING COST AND QUANTITY OF METAL

FOR MAKING OF ONE SET CUTTING TOOL

weight  cost

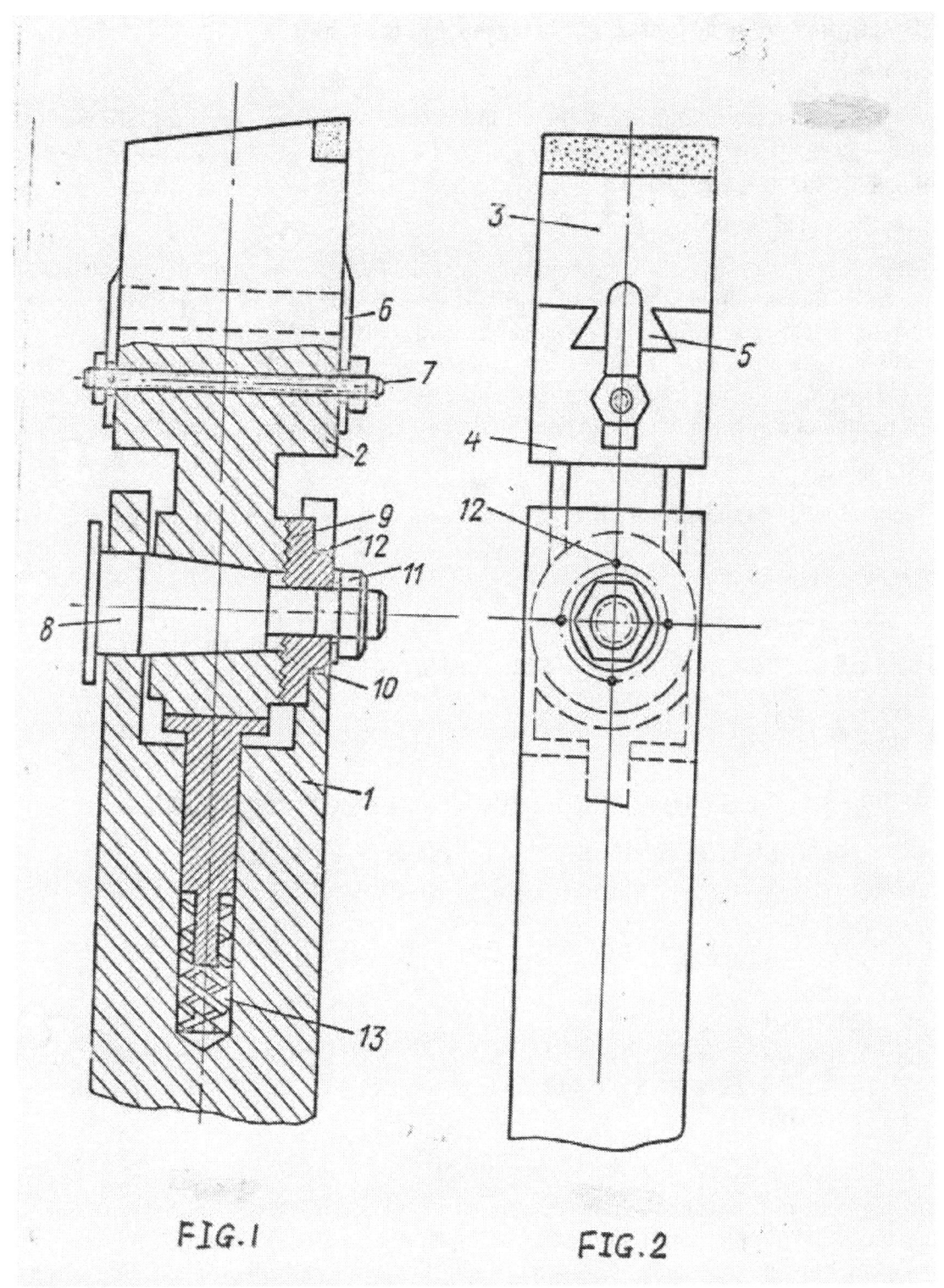

FIG.1   FIG.2

LATHE TOOL

Description of the advantages of invention "Lathe tool"

Usefulness:

This invention is used to expand the technological possibility of lathe tool for machining or boring the different materials into account that cutting tool changes the side and end cutting edge angles.

Novelty and profitable:

The present cutting tools for machining have many different right-hand and left-hand single-point tool which are applied for turning, planing, boring and other cutting processes and having integrity shank and nose for each type of tool.

This invention removes these defects so that tool is made in view separate head (nose) with the cutting edge angles which is fixed in shank of said lathe tool on conical axis so that last is tighten in conical hole of head and fixed in given position by means of screw.

Profitable:

General value of this invention is the decreasing of quantity of metal for making up the cutting tools and cost of job, and also this invention could expand the technological possibility of the machining processes in whole.

8. Pipeline for pneumatic transportation of granular materials

UNION OF SOVIET SOCIALIST REPUBLICS

(19)   **SU (11) 1164174 A**
(51) 4 B 65 G 53/52

USSR STATE COMMITTEE OF INVENTIONS AND DISCOVERIES

SPECIFICATION TO AUTHOR'S CERTIFICATE

(21) 3501770/27-11
(22) 15.10.82
(46) 30.06.85 Bull.#24
(72) A.I.Rozenblat
(53) 621.867.8 (088.8)
(56) Author's Certificate USSR # 816913 cl.B 65 G 53/52, 1979
(54) PIPELINE FOR PNEUMATIC TRANSPORTATION OF GRANULAR MATERIALS.

ABSTRACT

(57) Pipeline for pneumatic transportation of granular materials, comprising straight and curve pipes, with equal through passage sections and the fixture installed between it for twisting of flow granular material distinguishing that with objective of reducing of power consumption on transportation, said fixture for twisting of flow granular material presents in view of straight branch pipe which is fulfilled with equal as same pipe through passage section and arranged with possibility of rotation from a drive so that axis of rotation displaced accordingly longitudinal of axis of branch pipe.

1 CLAIM; 2 DRAWING FIGURES

This invention is designed to decrease power consumption at the transportation of different granular materials by pneumatic method also gives the possibility of decreasing the wear on a pipeline, particularly on such elements as elbows.

Pipelines for pneumatic transportation of granular materials (sand, ash, and other abrasive materials) have short periods of durability, particularly of elbows. Pneumatically moving granular materials through a pipeline presents particular problems wherever the pipeline must bend. At these elbows in the pipeline, several things occur.

First, particles that have been travelling in a straight line through the pipe strike the wall of the elbow with force sufficient to cause constant wearing at that point. Thus elbows tend to wear out quickly, and replacement is costly.

Second, the movement of the granular substance naturally slows down at the curves in its path. The resistance causes a loss of pressure, inhibiting the free movement of material through the elbow. Consequently, the product does not flow with equal speed and ease through all segments of the pipeline. It flows quickly through straight lines of pipe but tends to back up at the elbows.

Third, getting granular materials to pass at consistent speed and volume through elbows requires increased power to be applied.

In Figure 1 shown suggested pipe, in cross-section; in Figure 2—section A-A on Figure 1. The pipeline consists of the straight pipe 1, which is made with the special flange 2. The flange 2 provides a landing place 3 for bearings 4 and bushing 5. The branch pipe 6 has the flanges 7 and 8. A landing place 9, is constructed on the flange 8 and on another flange 7 there is a landing place 10 for bearing 11 and 12. The landing place 13 is completed in the elbow 14. Also on the second flange 7, the gears 15 are placed (or assembled by a forced fit of a bushing 16 on which there are teeth 15).

The ring 17 arranges on the parts separated from the bronze, and on the bushing 5 assemblies the fixture for the twisting of the flow is in view of the straight branch pipe 6. On the other side of branch 6 in a radial thrust bearing 12, there is a bushing 18, and at another bearing 12 a landing place 9 is fixed. After this, a bearing 20 inserts into the flange 19 of the elbow 14, and the process of assembling makes, together with the branch pipe 6 the following:

The persistent ring 21 input fixes the elbow 14 on the bushing 18, and under the loaded strength of the pipe 1 and the elbow 14, all system fixes. The branch pipe 6 rotates in during work around an axis from the drive 22 because of the bearings 4,11 and 12,20.

The rotation of the branch pipe 6 provided by the motion of a handle drive 23, which fixes the pipe in a given position and also disconnects the system 24 from the drive 22.

The working process of the invention is as follows:

The gas—material mixture (for instance, abrasive material and air) comes to the pipe 1 and then to the branch pipe 6. The pipe 6 rotates constantly under the action of the drive 22. The branch pipe 6 rotates with an angular rate equal to the period of rotation and moves the gas-material mixture on a spiral or helical trajectory. At this point, a mixture falls on the convex inner surface of the pipe 14 at a non-definite location, and because of the displacement of the axis turning relative to the longitudinal axis of the branch pipe 6, the mixture uniformly uses this surface and removes local abrasive wear at a given point.

Besides this changing of the trajectory of flight, the moving particles in pipe 1 positively influences the use of the inner surface of the pipe 14 and the wear resistance of system increases considerably. The branch pipe 6 could be rotated periodically by the handle.

This means that at some time (calculated by experiment) the branch pipe 6-according to the displacement of the pipe 14 on a define angle, and after some period of time with the objective of equally wearing the outer (convex) side of the pipe 14-turns on the other angle until the turning of branch pipe 6 finishes on a 360-degree angle.

This operation slows the wear of the pipeline, and the cycle repeats again. The suggested invention gives the possibility of decreasing the expenditures of power consumption by including the fixture for the twisting of flow which also gives the possibility of decreasing the losses of pressure and wear on the elbow.

So, this unique, patented mechanical design for this pneumatic system causes the spiraling motion that propels materials smoothly around corners. Not only does this system reduce wear at the elbow, it also reduces the power consumption required to keep the products moving along. Efficiency of the overall flow through the line increases substantially.

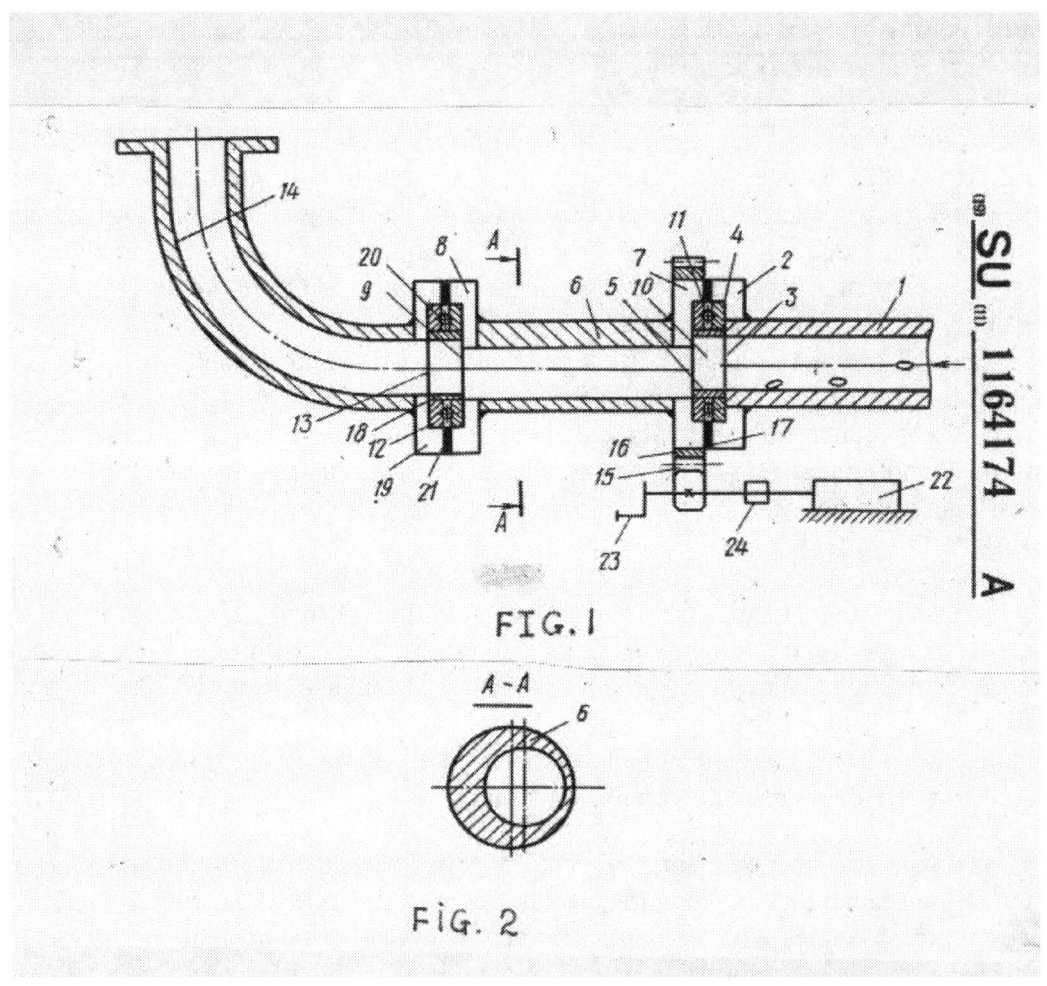

FIG.1

A-A

FIG. 2

Pipeline for pneumatic transportation of granular materials

32

8. PIPELINE FOR PNEUMATIC TRANSPORTATION OF GRANULAR MATERIALS

NON-CONFIDENTIAL INVENTION SUMMARY

AIR #298

PIPELINE FOR PNEUMATIC TRANSPORTATION OF GRANULAR MATERIALS
(Inventor: Anatoly Rozenblat)

INVENTION DESCRIPTION:

Device is used to reduce of power consumption on the transportation different granular materials by pneumatic method and also to decrease the losses of pressure and wear of pipeline particularly the elbows of these systems.

UTILITY AND ADVANTAGES:

Pipeline for pneumatic transportation of granular materials (sand, ash, and other abrasive materials) has short period of durability particularly of elbow.

This invention removes these defects so that between straight and elbow are installed with equal through passage sections additionally the fixture for twisting of movable granular material so that said fixture also is fulfilled with equal the same through passage section and arranged with possibility of rotation from drive around of this fixture in view of branch pipe.

PATENT AND DEVELOPMENT STATUS:

This invention has been patented in the Soviet Union; the inventor has brought the technology and know-how when he immigrated to the USA.

FURTHER INFORMATION:

Copies of the patent are available upon request. Data demonstrating its efficacy of use are available from the inventor under cover of a Confidential Disclosure Agreement. Commercial rights are available for exclusive, non-exclusive or field-of-use exclusive licensing. When making inquiries PLEASE REFER TO THE AIR # SEEN ABOVE.

9. Method of processing curvilinear channels of A.I.Rozenblat

UNION OF SOVIET SOCIALIST REPUBLICS

(19) SU **(11)** 1441658 A1
(51) 4 B 24 B 39/02

USSR STATE COMMITTEE ON INVENTIONS AND DISCOVERIES

SPECIFICATION TO AUTHOR'S CERTIFICATE

(21) 3823818/25-27
(22) 12.11.84
(75) A.I.Rozenblat
(53) 621.923.77 (088.8)

(56) Yu.G.Proskuryakov. Hardening Calibrating Methods of Processing, Moscow, Mashinostroyeniye,1965, p.90.

(54) METHOD OF PROCESSING CURVILINEAR CHANNELS OF A.I.ROZENBLAT

(57) This invention is related to mechanical processing of curvilinear channels, namely to finishing out. The objective is to increase efficiency. For this purpose, working tools-balls are filled into the curvilinear channel.

Then, separate work pieces are assembled into a set in such a way that the end of a channel of each preceding tap coincides with the beginning of a channel of each next one, and all together they form a closed curvilinear space for tools.

Concavities of all taps are turned toward the center. The balls are charged. Upon rotation, centrifugal forces press the balls to the surface of the channels, and the start to move toward the center. This makes it possible to intensify the processing.

Description

The method of processing the curvilinear channels is related to the mechanical processing of channels predominantly in the elbow of systems for pneumatic and hydraulic trunk lines transporting the ferromagnetic materials.

This method and device shown on Figure 1, is used to increase the efficiency of mechanical processing of the curvilinear channels, namely to finishing out and also to decreasing abrasion.

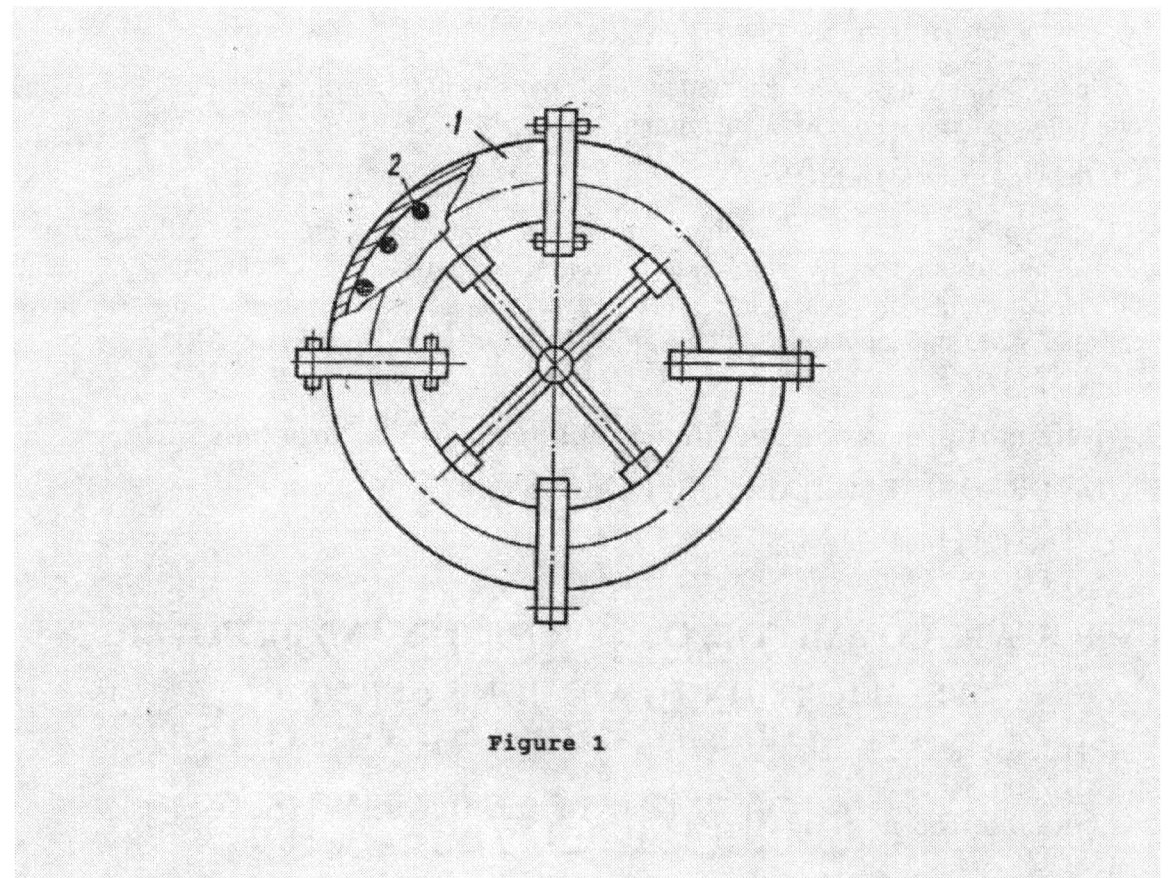

Figure 1

METHOD OF PROCESSING CURVILINEAR CHANNELS OF A.I.ROZENBLAT

So, the device for cutting inner surfaces of the channels bends includes such elements as working system 1 and tool-balls elements 2 which are filled into a work piece system 1 with inner curvilinear channel.

The diameter of the balls is smaller than that of the channel, and their firmness is determined by the technological conditions.

Several separate work pieces 1 are assembled into a set in such a way that the end of the channel of each preceding tap coincides with the beginning of the channel of the next one, and all together the form a closed curvilinear space for placing operating tools, and the concavities of all taps are turned to the center.

The assembled set is set into rotation around its central axis. The internal channel of the connected taps may form a trajectory approaching a circle. Upon rotation, centrifugal forces press the balls 2 to the surface of the channels and start to move them to the center and over the body, conducting the processing.

The balls can be charged or coated with an abrasive material which intensifies processing. The processing can be also performed with supply of lubricant coolant.

CLAIM:

1. A method of processing curvilinear channels, predominantly taps of systems of pneumatic and hydraulic trunk lines, which includes placing a working tool into the channel of taps and its application to the walls of the channel upon relative transfer of the working tool and the walls of the channel.

    The method is distinguished by the fact that in order to increase efficiency, the taps are series connected to each other in such a way that the end of the channel of each preceding tap coincides with the beginning of the channel of each next one, and all channels form a closed curvilinear space for working tools, and concavities of all taps are turned to the center. In addition, the relative transfer of the working tool and the walls of the channel results from rotation of interconnected taps.

2. The method according to item 1 is distinguished by the fact that the balls are used as the working tool.

3. The method according to items 1 and 2 is distinguished by the fact that the internal channel of the connected taps forms a trajectory, approaching a circle.

NON-CONFIDENTIAL INVENTION SUMMARY

AIR #294

METHOD OF PROCESSING CURVILINEAR CHANNELS OF A.ROZENBLAT
(Inventor: Anatoly Rozenblat)

INVENTION DESRIPTION:

This device is used to increase efficiency mechanical processing of curvilinear channels, namely to finishing out and also to decrease abrasion.

UTILITY AND ADVANTAGES:

This method is related to mechanical processing of curvilinear channels predominantly in elbow systems for the pneumatic and hydraulic trunk lines transporting ferromagnetic material.

The present methods and arrangement do not guarantee the good quality of inner surface of elbow in manufacturing processes. And for this reason the inner surface has rough profile and naturally this defect promotes to the abrasive increasing of said inner surface for the elbow.

This invention increases resistance and efficiency of curvilinear channels such as elbow particularly in pneumatic systems.

The separate working pieces of elbows are assembled into one set in such a way the end of a channel of each preceding elbow coincides with the beginning of a channel of each next one, and all together they form a closed curvilinear space for tools and then the working tools-balls are filled into the curvilinear channel and balls are charged.

Upon rotation, the centrifugal forces press the balls to the surface of the channels, and they start to move toward the center. This makes it possible to intensify the processing.

PATENT AND DEVELOPMENT STATUS:

This invention has been patented in the Soviet Union; the inventor has brought the technology and know-how when he immigrated to the USA.

FURTHER INFORMATION:

Copies of the patent are available upon request. Data demonstrating its efficacy of use available from the inventor under cover of a Confidential Disclosure Agreement. Commercial rights are available for exclusive, non-exclusive or field-of-use exclusive licensing. When making inquires, PLEASE REFER TO THE AIR # SEEN ABOVE.

10. Multi-positional combined die of A.I.Rozenblat for cutting of sheet materials

UNION OF SOVIET SOCIALIST REPUBLICS

(19) SU **(11)** 1500416
(51) 4 B21 D 37/08,28/14

USSR STATE COMMITTEE OF INVENTIONS AND DISCOVERIES

SPECIFICATION TO AUTHOR'S CERTIFICATE

(21)   4354571 /25-27
(22)   08.12.87
(46) 15.08.89 Bull.#30
(75) A.I.Rozenblat
(53)   621.961.2 (088.8)
(55)   Patent USA #3823630 cl. B26 F 1/ 02, published 18.11.74
(54)   MULTI-POSITIONAL COMBINED DIE OF A.I.ROZENBLAT FOR CUTTING OF SHEET MATERIALS

ABSTRACT

(56) Invention relates to the plastic metal working process, particularly to the die and punching processes.

Objective of this invention is the decreasing of labor input and improvement of operation conditions by means of exception of transferring of blank sheet from one position to another.

Die contains the plates 2 and 4, punches 1,6,8 and die 5,7,3. Between of these pair of punch 1 and die 5, of punch 6 and die 7, of punch 8 and die 3 are situated accordingly the elastic elements 9, 10 and 11. Stiffness of element 11 more stiffness of element 10, and stiffness of element 10 more stiffness of element 9. Dies 5 and 7 are joined with the plates 2,4 and also with hollow columns 13.

In these columns 13 are installed the hydraulic ram 16 with stops 17 and 18. The spaces 14 in columns 13 are connected with the source of pressure, for instance hydraulic. Punches 6 and 8 are tied with dies 5 and 7 so that above-named punches move in plane of joint with dies when the sheet metal blank is moving from one position to another. The metal strip 32 feeds under punch 1 when makes the punching of detail said punch 1 raises up and under its again feeds the following section of metal strip 32. At this time the punch 6 removes in side so that stamped metal blank falls down through on the surface of die 7.

Punches 1 and 6 make the following operation processes and then punch 1 again ascend in normal upper position. In this period the punches 6 and 8 remove in side so that stamped metal blanks fall down through on the surfaces of dies 7 and 3. Under punch 1 again feeds the metal strip 32. Further cycle repeats again.

3 CLAIMS; 3 DRAWING FIGURES.

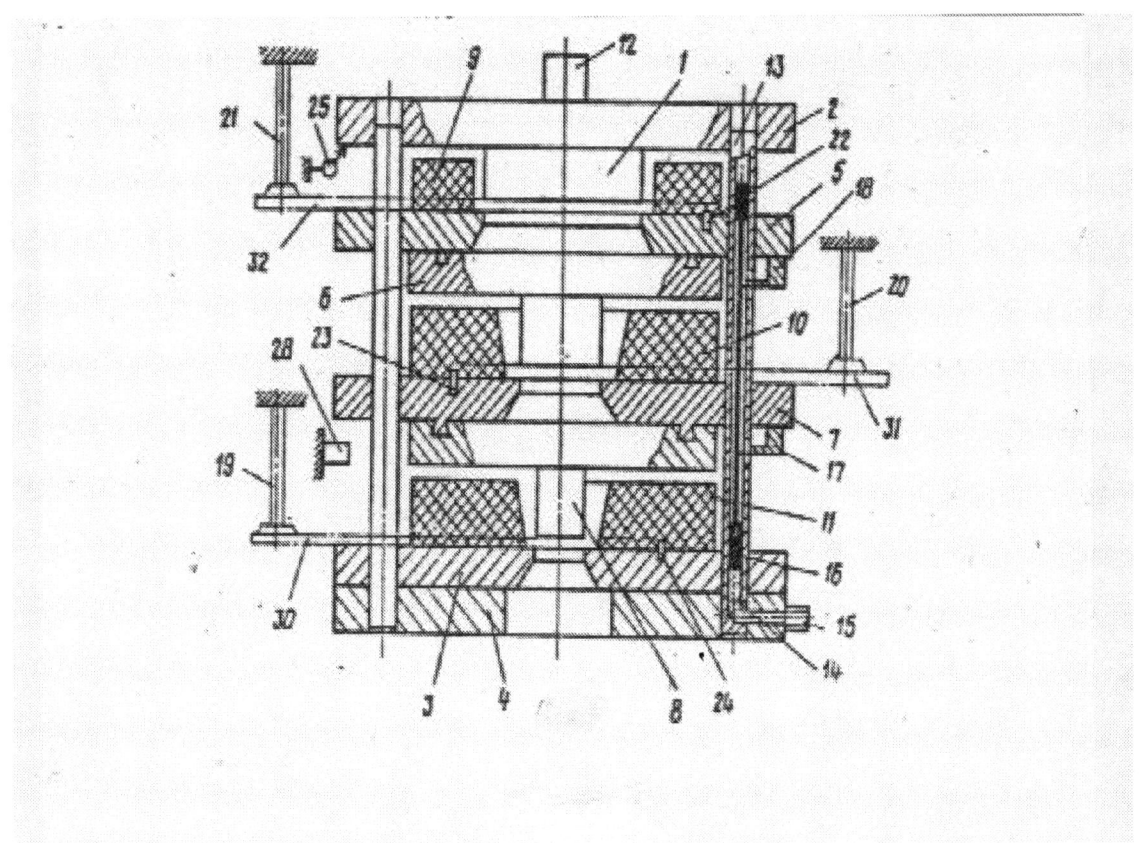

Description

In flexible production, a recommended device to use is the multi-operational combined die of A.I.Rozenblat as shown on Figure 1, the die in general view; on Figure 2—block—diagram of automatic work with independent feed of material and on Figure 3-block-diagram of automatic work with feed of material from upper position.

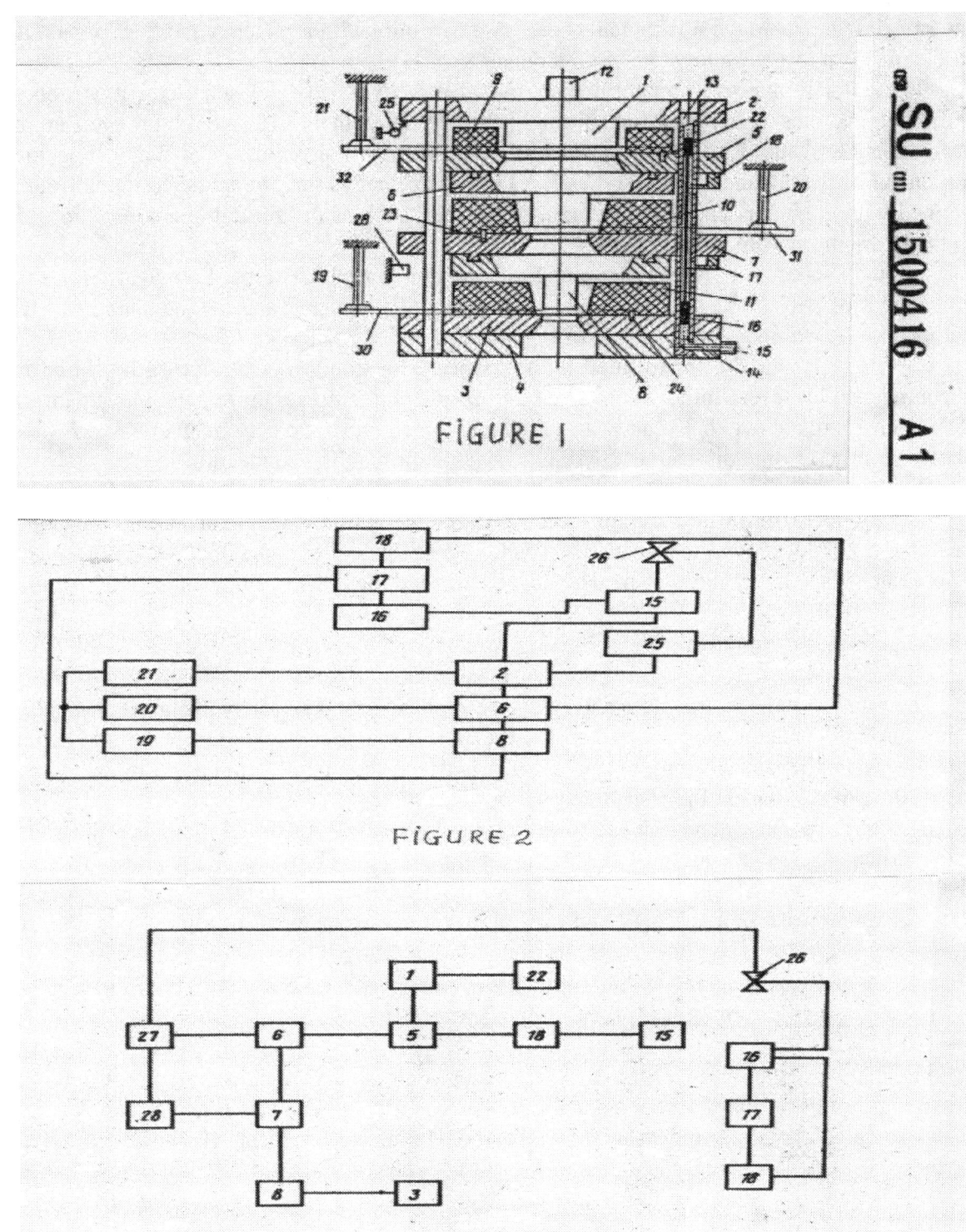

FIGURE 1

FIGURE 2

FIGURE 3

MULTI-OPERATIONAL COMBINED DIE OF A.I.ROZENBLAT FOR CUTTING OF SHEET MATERIALS

This die contains the upper punch 1 fixed on a plate 2 and also the bottom die 3 fixed on a plate 4. In the middle part of the stamp between the punch 1 and the die 3, an additional intermediate movable die 5 is joined to the "swallow tail" with a punch 6 and die 7 which is also connected to a "swallow tail" with a punch 8.

On the upper plane of the intermediate movable die 5 is a fixedelastic element, a rubber buffer 9. The dies 7 and 3 and accordingly the elastic elements 10 and 11 have different stiffnesses. A shank 12 fixes the plate on the press with the objective of controlling the alignment of the punches 1,68 and dies 5,7,3 and also ensures that the guide columns 13 are aligned with the inner spaces 14 for input of material such as oil through the connection 15 to the hydraulic system.

Through spaces 14, move the plunger 16 with controlled stops 17 and 18 for the adjustment of two of the punches 8 and 6 with two of the dies 7 and 5. The stamp can be equipped with manipulators 19,20 and 21 for keeping of fed material.

The stops 17 and 18 are fulfilled by accessory. The fixing of material in the stamp is accomplished by several other stops 22,23,24. To guarantee the control of the movement of the punch 1 a microswitch 25 can be installed and joined with an electromagnet valve 26(not shown) for the feeding oil from the hydraulic system through the connection 15.

Displacement of the punches 6 and 8 in the plane perpendicular to the axis supplies pressure by interacting with the transducer 27,28 (not shown), which is joined with the valve 26. A stop 29 (not shown) is provided to keep the dies 7 and 5 in place in case of an absence of pressure in the hydraulic system. The blanks strips adopted the positions shown 30,31 and 32.

- A stamp sets up and works the following way:

The stamp installs on the press and fixes to crosshead (not shown) by means of shank 12. Later through connection 15 moves a pressure. The distance between punches 6 and 8 and dies support by means of stops 18 and 17, and between the punch 1 and die 5 by means of controlling of crosshead of press.

Between of punches abd dies are installed the sheet-strips 30,31 and 32. When the plate move down successive cutting of sheet strips making on the above-named position because the elastic elements 9,10,11 have the different stiffness. After of finishing of cutting process again gives a pressure in space 14 and take place the return of tools in primary position by means of plunger 16. The plate 2 with the punch 1 again moves up by means of press.

- Action of block-diagram at the automatical work, with independent sheet material is given layerly (Figure 2)makes the following way:

The plate 2 coming down acts upon on microswitch 25 which gives signal on the valve 26 so that in this period pressure does not give into space 14. Punches 6 and 8 and dies 5,7 lower down before contacting with strips 30 and 31 and elastic elements 10 and 11. Moving further above-named punches and dies make consistentical cutting of blank.

When punch 1 returns in upper position liquid comes in space 14 and the other tools return in normal position.

- Action of block diagram at automatical work, when material gives in zone of cutting from upper position (Figure 3) makes the following way:

Pressure comes in space 14 so that punch 1 makes punching process and then this punch 1 returns in upper position and in this period transducer 27 receives impulse and punch 6 moves in side to the guides of type "swallow tail."

Made blank drops to the die 7 and new blank gives under punch. The punch 6 takes a working position. When punch 1 moves down in punching process, the punch 6 also does this process.

And then punch 1 again moves up and this period transducer 27 and 28 receive impulse and punches 6 and 8 move in side. In this moment made blanks drop down accordingly the above-named dies. And the process of punching repeats further.

Introduction of this invention will allow to decrease labor input and to improve the conditions of operating process by way of exception of transferring of blank with one position to the another automatic work with moving of sheet from upper position.

CLAIM:

1. Multi-positional combined die for cutting of sheet materials comprising at least way two pair punches and dies for divisional operations, which are installed to the axis of pressure of die so that one pair of punches installed in movable of part die and another pair installed in unmovable part and the other said die and punch connected between them in total unit and installed between movable and unmovable parts of die and also having elastic elements which from it situated between corresponding pair punch-die *distinguishing* so that with objective of decreasing of labor input and improvement condition of operating by way of exception of transferring of blank with one position to the another, said die supplied by guides elements joining die from unit with movable and unmovable of parts of die and punch connected with this die for possibility of moving in plate perpendicular of axis of pressure of die.

2. Multi-positional combined die on claim 1 *distinguishing* so that guides elements fulfilled in view of hollowness columns and equipped by plungers which are installed at these hollowness space and joined with unit so that under plunger parts of these spaces connected with source of pressure.

3. Multi-operational combined die on claim 1 *distinguishing* so that each the following elastic element in direction from unmovable part to the movable part is fulfilled with smaller stiffness that preceding element.

11. Transport pipeline

UNION OF SOVIET SOCIALIST REPUBLICS

(19) SU (11) 1134504 A
4(51) B 65 G 53/52

USSR STATE COMMITTEE ON INVENTIONS AND DISCOVERIES

SPECIFICATION OF AUTHOR'S CERTIFICATE

(21)   3451278/27-11
(22)   06.08.82
(46) 01.15.85 Bulletin №2
(72) A.I.Rozenblat
(53) 621.867 (088.8)
(57)   1.USSR Author's Certificate №816913, Cl.B65 G 52/53,1981

(55)   **(57)** TRANSPORT PIPELINE which comprises a horizontal and vertical pipes, a bend located between them, and an element for changing trajectory of motion a flow of loose materials installed in the end of the horizontal pipe before the bend.

The pipeline is distinguished by the fact that in order to simplify design and to improve operating convenience, the element for changing trajectory of motion of a flow of loose materials is made in the form of an elastic insert fastened between the faces of the horizontal pipe and the bend with the possibility of closing the zone of joint of the bend and the horizontal pipe below the horizontal diametric plane of the pipe.

In addition, the upper surface of the insert from the side of the horizontal pipe is conjugated with its internal surface, and the other portion of the insert is located at the acute angle to the internal surface of the bend.

The invention is related to pipeline transport and can be used in different branches of industry for pneumatic and hydraulic transport of loose materials.

A transport pipeline is known which comprises a horizontal and vertical pipes, a bend located between them, and an element for changing trajectory of motion of a flow of loose materials installed in the end of the horizontal pipe before the bend 1.

Faults of the known pipeline include its design complexity and instability of operation of the spiral-shaped element installed freely in the horizontal pipe before the bend.

The aim of the invention is to simplify design and to improve operating convenience. The formulated goal is achieved by the fact that in the pipeline, containing the horizontal and vertical pipes the bend, located between them and the element for changing trajectory of motion of flow of loose materials installed in the end of the horizontal pipe before the bend, the element for changing trajectory of motion of loose materials is made in the form of a flexible elastic insert fastened between the faces of the horizontal pipe and the bend with the possibility of closing the zone of joint of the bend and the horizontal pipe below the horizontal diametric plane of the pipe.

In addition, the upper surface of the portion of the insert from the side of the horizontal pipe is conjugated with its internal surface, and the other portion of the insert is located at the acute angle to the internal surface of the bend.

Figure 1 shows general view of the transport pipeline; Figure 2 shows the same A-A section of Figure 1 (from the side of the entry connection pipe).

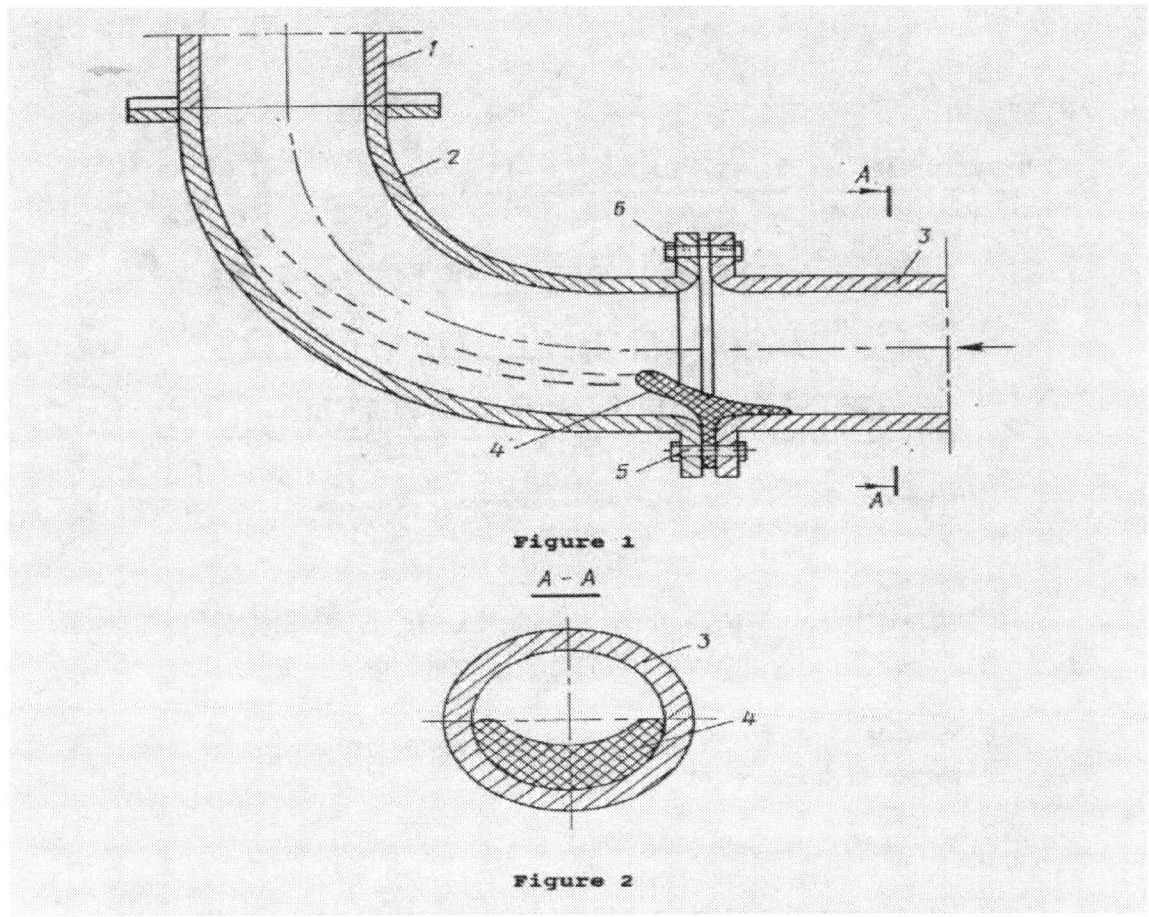

Figure 1

A - A

Figure 2

TRANSPORT PIPELINE

The transport pipeline contains an exit connection pipe 1, a bend 2, an entry connection pipe 3, an elastic insert 4, a bolt 5, a nut 6. The insert has surface conjugated with the internal surface of the horizontal pipe and oriented toward the bend at acute angle, the insert is located in the site of joint of the horizontal pipe and the bend.

The transport pipeline operates in the following manner:

In the process of motion of a flow of loose materials through the entry connection pipe 3, particles of material, reaching the elastic insert 4, change trajectory of motion which is characterized by different slopes depending on weight and rate of motion of particles. As a result of scattering of particles, wear of the bend is not local but rather takes place over a certain area. Here, rigid fastening of the elastic element creates conditions for an increase in stability of operation of the transport pipeline.

The invention also helps to simplify the design of the element which changes trajectory of motion of a flow.

TECHNOLOGY TARGETING, INCORPORATED

A Non-Profit Scientific/Educational Organization

2940 South Warr
Salt Lake City, UT 84109
(801) 487-9800

NON-CONFIDENTIAL INVENTION SUMMARY
[Information Provided Without Obligation]

TTI# - 292
INTERNALLY-MODIFIABLE PIPELINE TRANSPORT
[Inventor: Anatoly Rozenblat]

INVENTION DESCRIPTION: Device to decrease abrasion of pipeline surfaces by transported materials.

UTILITY AND ADVANTAGES: Pipelines are used for transport of a variety of materials, including solid materials in a liquid suspension. Such materials cause continual wear and damage to the pipeline which transports it, due to its constant contact with the inside of the pipe. This damage is most pronounced at the elbows and curves of the pipeline, when the differential inertia of these materials sends them against the pipeline wall.

This invention utilizes a uniquely designed insert at such bends and elbows. This insert alters the trajectory of the suspension as it enters the curve, converting it to a more homogeneous stream with reduced separation of the abrasive materials. This modified trajectory has been shown to materially reduce the abrasion to the pipeline.

PATENT AND DEVELOPMENT STATUS: This invention has been patented in the Soviet Union; the inventor brought the technology and know-how with him when he emigrated to the US.

FURTHER INFORMATION: Copies of the patent are available upon request. Data demonstrating its efficacy of use are available from the inventor under cover of a Confidential Disclosure Agreement. Commercial rights are available for exclusive, non-exclusive or field-of-use exclusive licensing. When making inquiries, PLEASE REFER TO THE TTI # SEEN ABOVE.

CHAPTER TWO
THE INNOVATIONS
IN USA

1. Multi-operational die with cone

Construction of multi-operational die with cone is shown in Figures 1,2 and 3 below.

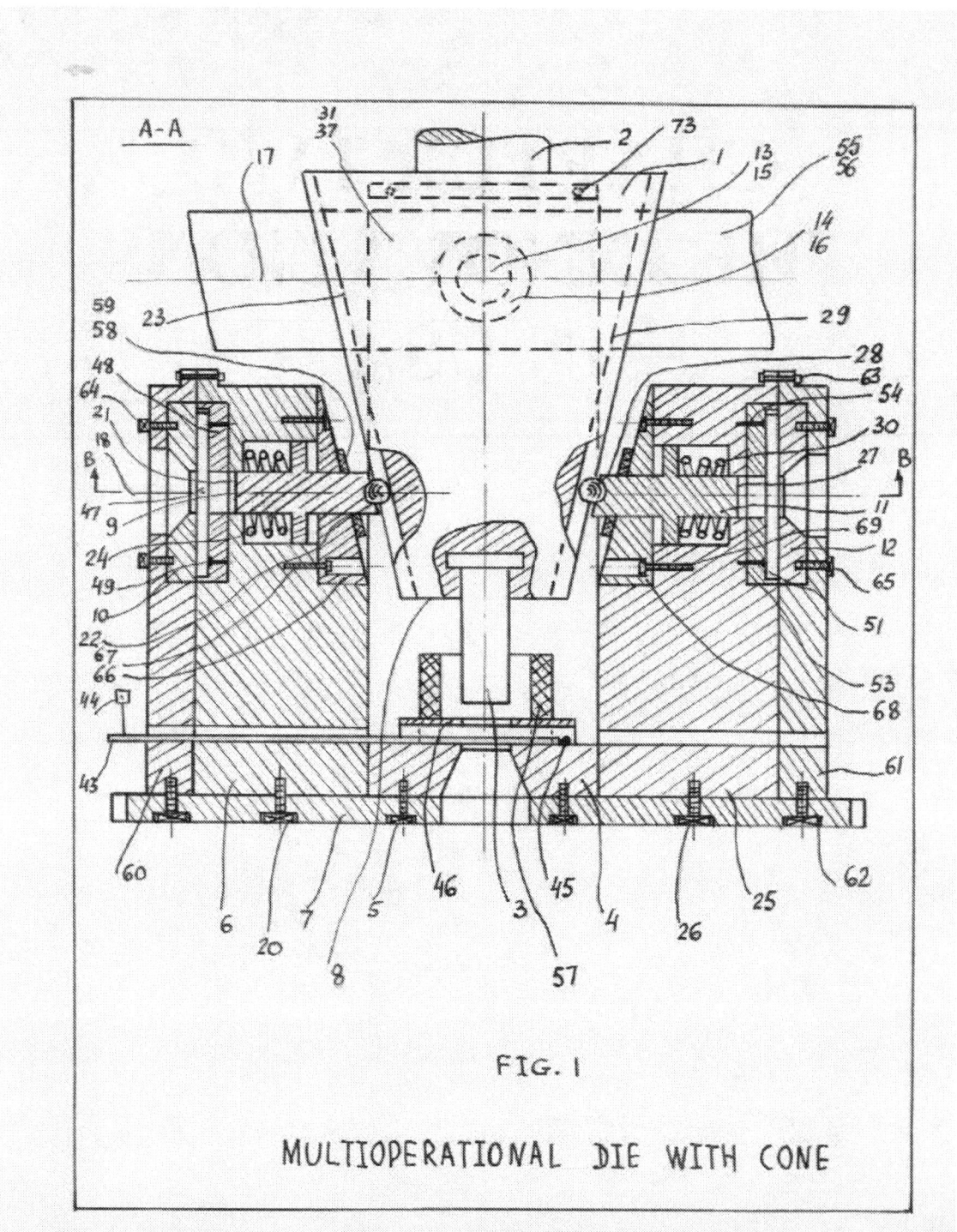

FIG. 1

MULTIOPERATIONAL DIE WITH CONE

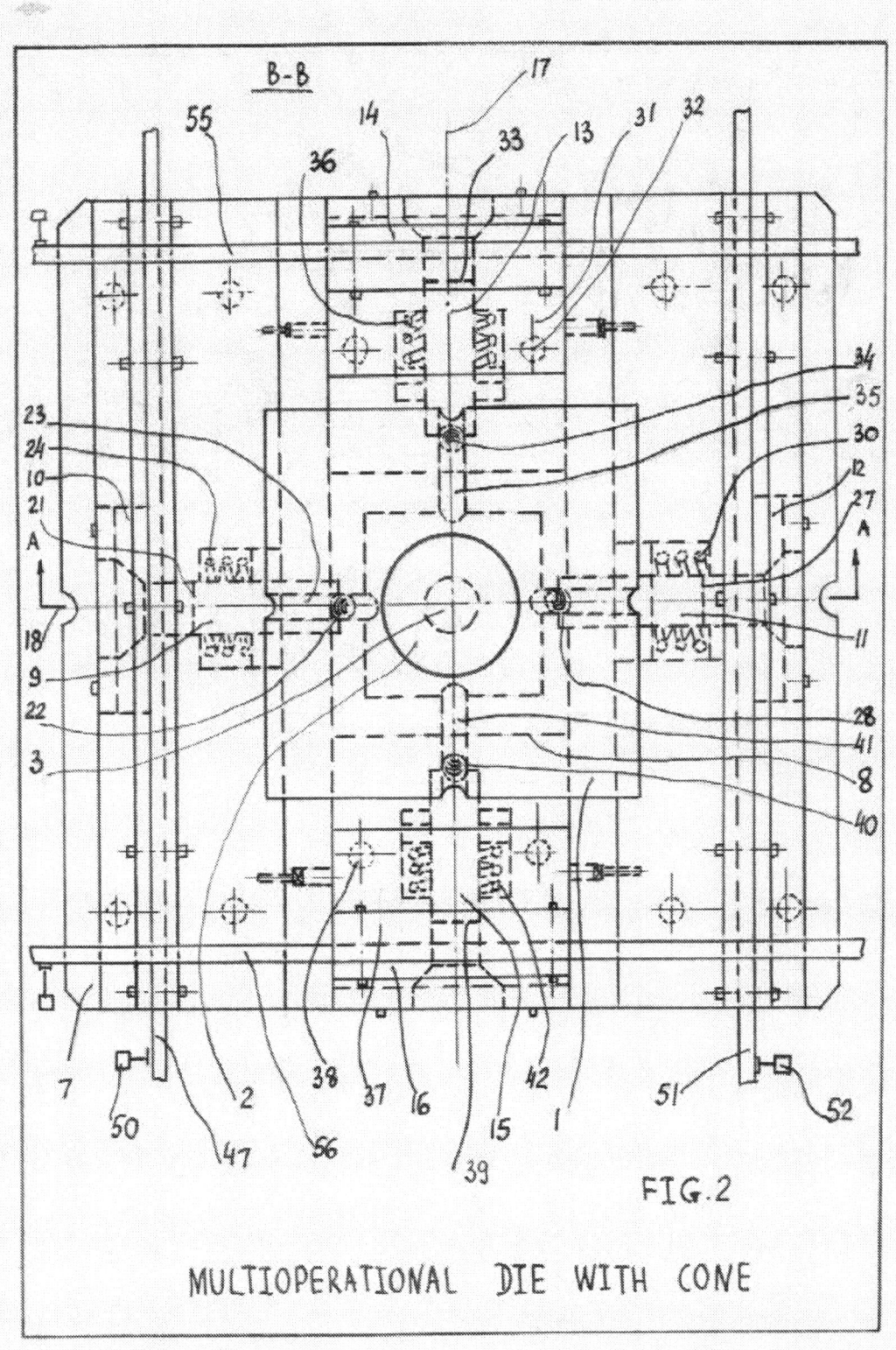

FIG.2

MULTIOPERATIONAL DIE WITH CONE

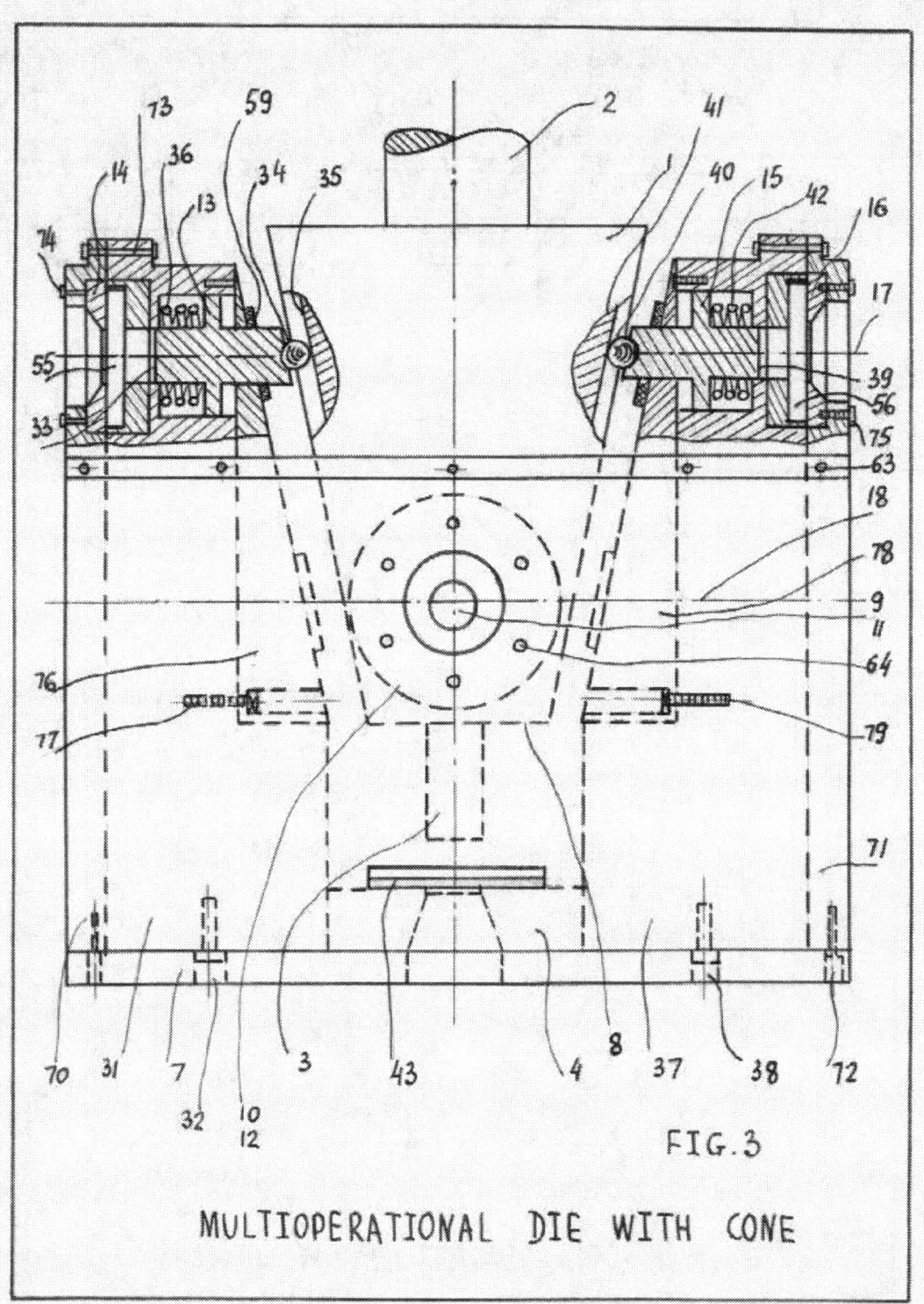

FIG.3

MULTIOPERATIONAL DIE WITH CONE

The die with cone includes such elements as the movable punch holder 1 which is fixed on one side with the cylindrical shank 2 and on the other side is joined with the vertical fixed elements of a punch 3 and a die 4. The die 4 is fixed by the screw elements 5 to the body 6 and to the lower plate 7.

The peculiarities of this stamp is that the movable punch holder 1 is in view of a truncated pyramid so that the lower base 8 is situated to the side of the main punch 3. Also the automated universal stamp includes such elements as two pair of horizontally and diametrically situated elements of punch 9 and die 10 and a second punch 11 and die 12.

The other pair includes such elements as a punch 13 and die14, and a punch 15 and die 16 which effect the working technological process of stamping. Thus, the one pair of elements—punch 13 and die 14 and also punch 15 and die 16-is displaced with axis 17 relative to the other pair of elements-punch 9, die 10 and punch 11, die 12-which have a total axis 18.

The peculiarities of the stamp consist in that each punch (9 and 11 of one pair and punches 13 and 15 of the other pair) are not fixed firmly with the movable punch holder 1. The punch 9 of the first system is fixed in the body 19 with the screw elements 20 and has a profile 21 such that there is a circle at one side while the other side of the punch has the bearing 22 which contacts the cone 23 of truncated pyramid.

The punch 9 of the first system has reciprocating motion: at the execution of the work process from the movable punch holder 1 and at the reverse motion (without load) because of the reverse spring 24 contacting the punch. The same picture has a place where the punch 11 of the first system is fixed in the body 25 by a screw 26 and has the profile 27 on one side aligning with the cutting work-piece (circle, ellipse, etc.) and on the other side the punch 11 has the bearing 28 with the body 29 attached to the adjoining side of truncated movable punch holder 1.

The punch 11 of the first system also has the reciprocating motion: at the execution of the work process from the movable punch holder 1 and at the reverse motion (without load) because of the reverse spring 30 contacting the punch 11.

The punch 13 on the other system is fixed in the body 31 by screws 32 and has the profile 33 on one side aligning with the cutting work piece (circle, ellipse, etc.) and on the other side the punch 13 has the bearing 34 which attaches the body 35 to the adjoining side of the truncated movable punch holder 1.

The punch 13 of the second system also has reciprocating motion: at the execution of the work process from the movable punch holder 1 and at the reverse motion (without load) because of the reverse spring 36 contacting this punch 13. The same picture with the punch 15 of the second system is fixed in the body 37 by screws 38 and has at one Side the profile 39 aligning with the cutting work-piece (circle, ellipse, etc.), and on the other side the punch 15 has the bearing 40 which attaches the body 41to the adjoining side of the truncated movable punch holder 1.

The punch 15 of the second system also has reciprocating motion: at the execution of the work process from the movable punch holder 1 and at the reverse motion (without load) because of the reverse spring 42 contacting with this punch 15.

- While the automatic universal stamp is in use, the following takes place:

The movable punch holder 1 installs separately in the press by using the cylindrical shank 2 (not shown) and then fastens the assembling die on the plate of the press and makes some field adjustments and finally fastens. After these operations, the sheet bar 43 moves by use of the universal manipulator 44 to the stop 45 of the window near the strip 46 and makes the cutting motion of the work-piece 43 by using the punch 3.

At the same time, the sheet bar 47 moves to the work area of the first system: the punch 9 and die 10 for the working process moves to the zone of the stop 48 which fastens with a screw 49. The sheet bar 47 moves by using the universal manipulator 50. At this point, the sheet bar 51 moves by use of the universal manipulator 52 to the zone of the first system: the punch 11 and die 12 for working process moves to the zone of the stop 53 which fastens with use of a screw 54.

Also, the sheet bar 55 moves to the zone of the second system; the punch 13 and die 14 and the sheet bar 56 move to the zone of the second system; and the punch 15 and die 16 move by use of universal manipulator (not shown) in the working processes. When the movable punch holder 1 moves down at this time, it makes the working process begin in the system of the first punch 3 and die 4 and also the working processes of the following systems:

The punch 9 and die 10 of the first system; the punch 11 and die 12 of the second system; as well as the other pairs of punches 13 and 15 and dies (14 and 16). In view of the fact that the punches 13 and 15 of the second system are situated above the first system, i.e the axis 17 relive to the axis 18 and the first system of punches 9 and 11 are displaced, then four of the sheet bars (55,56 and 47, 51) and also on the sheet bar 43 for the system (punch 15 and die 16) create the work processes as the movable punch holder 1 acts on each punch (3,9,11,13 and 15).

With the aim of safeguarding the work process and avoiding dangerous contact, the metal parts of the movable punch holder 1with the strip 46 have a special buffering system, such as rubber. The zone of contact of the bodies (6 and 25) with the body of the movable punch holder 1 has a buffering system 57, and the zone of contact of the other bodies (31 and 37) with the body of the movable punch holder 1 also has a buffering system (58 and 59).

With the aim of adaptability to manufacture and assembly of the first punch-die pair (9,19 and 11,12) to the bodies (6 and 25), the flanges (60 and 61) are adjusted which fastens the flanges to the lower part of the base 7 with screws 62 and on the upper part as well with screws 63.

Accordingly, the die 10 fastens in the flange 60 with screws 64 and the die 12 fastens in the flange 61 with screws 65. And besides the cover 66 is attached to the body 6 with the screws 67, and the body 25 is attached to the cover 68 with screws 69.

With the aim of adaptability to manufacture and assembly of the second punch-die pair (13,14 and 15,16) to the bodies (31 and 37), the flanges (70 and 71) are attached and fastened to the lower part of the base 7 with screws 72 and to the upper part with a screw 73. Accordingly, the flange 70 fastens to the die 14 with the screws 74, and the other flange 71 fastens to the die 16 with the screws 75. Also, the cover 76 is attached to the body 31 with screws 77 and other body 37 the cover 78 is attached with screws 79.

2. Telescopical cutting tool

The author of this innovation established experimentally some facts and designed the new Telescopical cutting tool which shown in Figures 1,2,3,4,5,6 and 7 and could be successfully used in manufacturing processes.

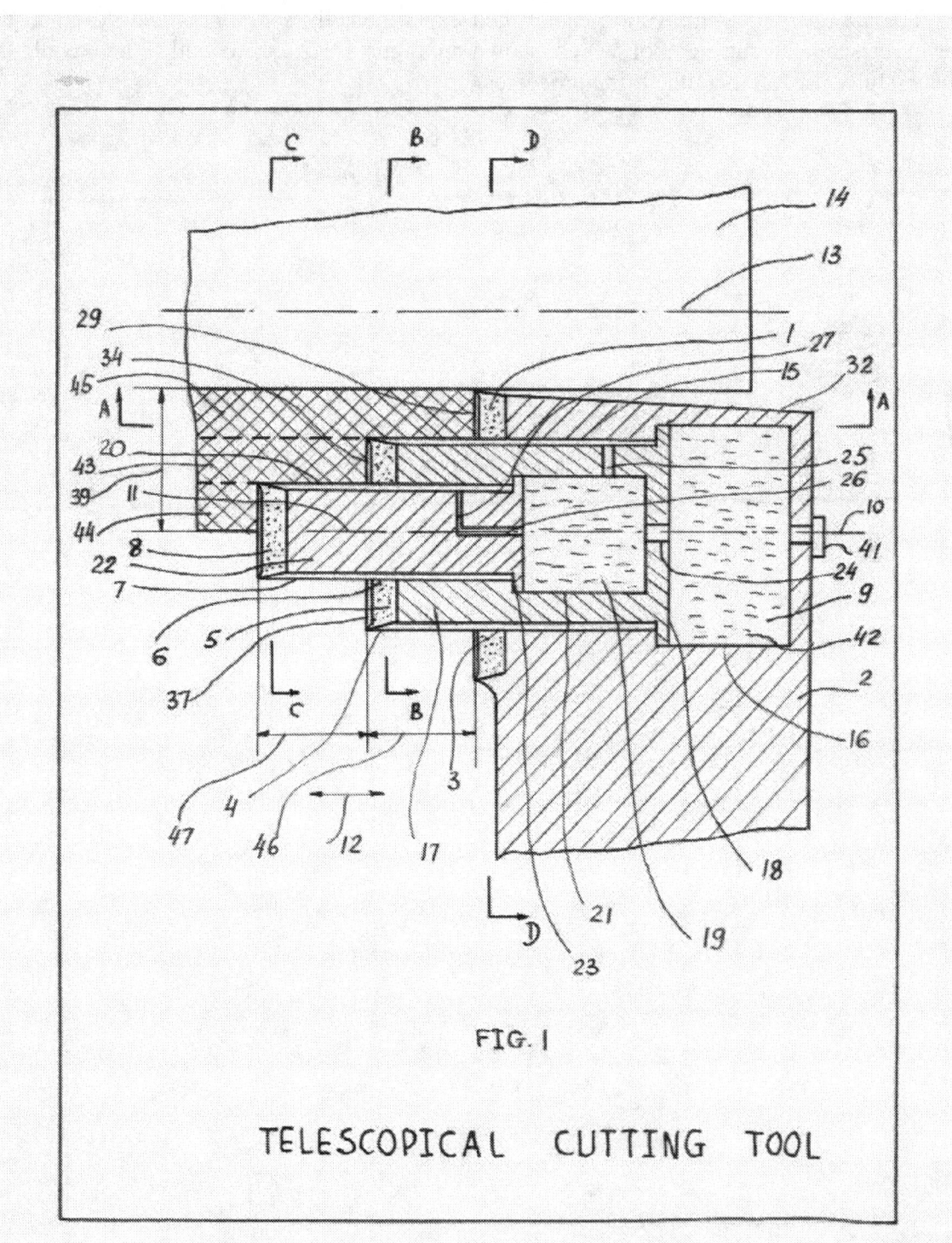

FIG. 1

TELESCOPICAL CUTTING TOOL

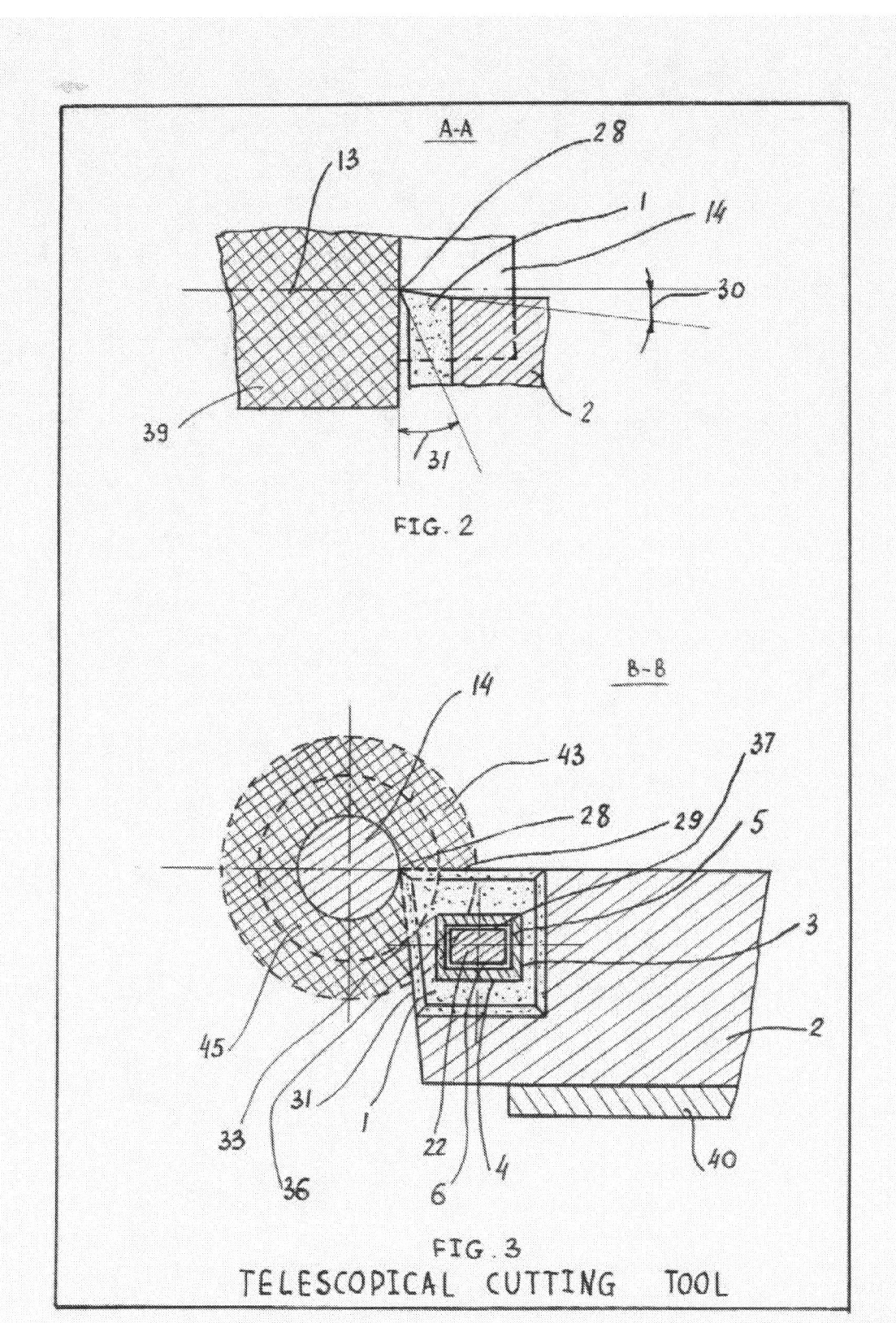

FIG. 2

FIG. 3

TELESCOPICAL CUTTING TOOL

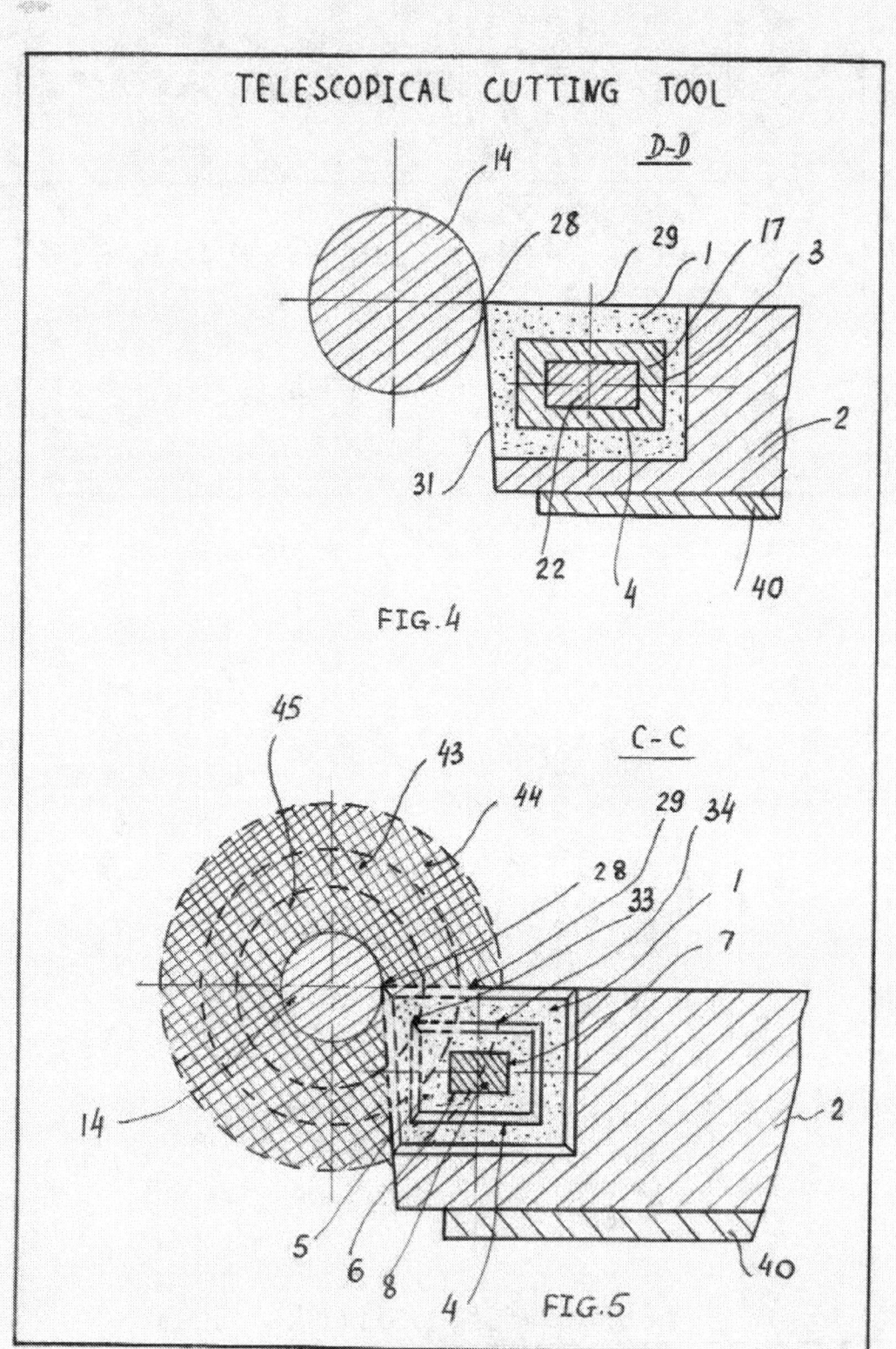

TELESCOPICAL CUTTING TOOL

FIG.4

FIG.5

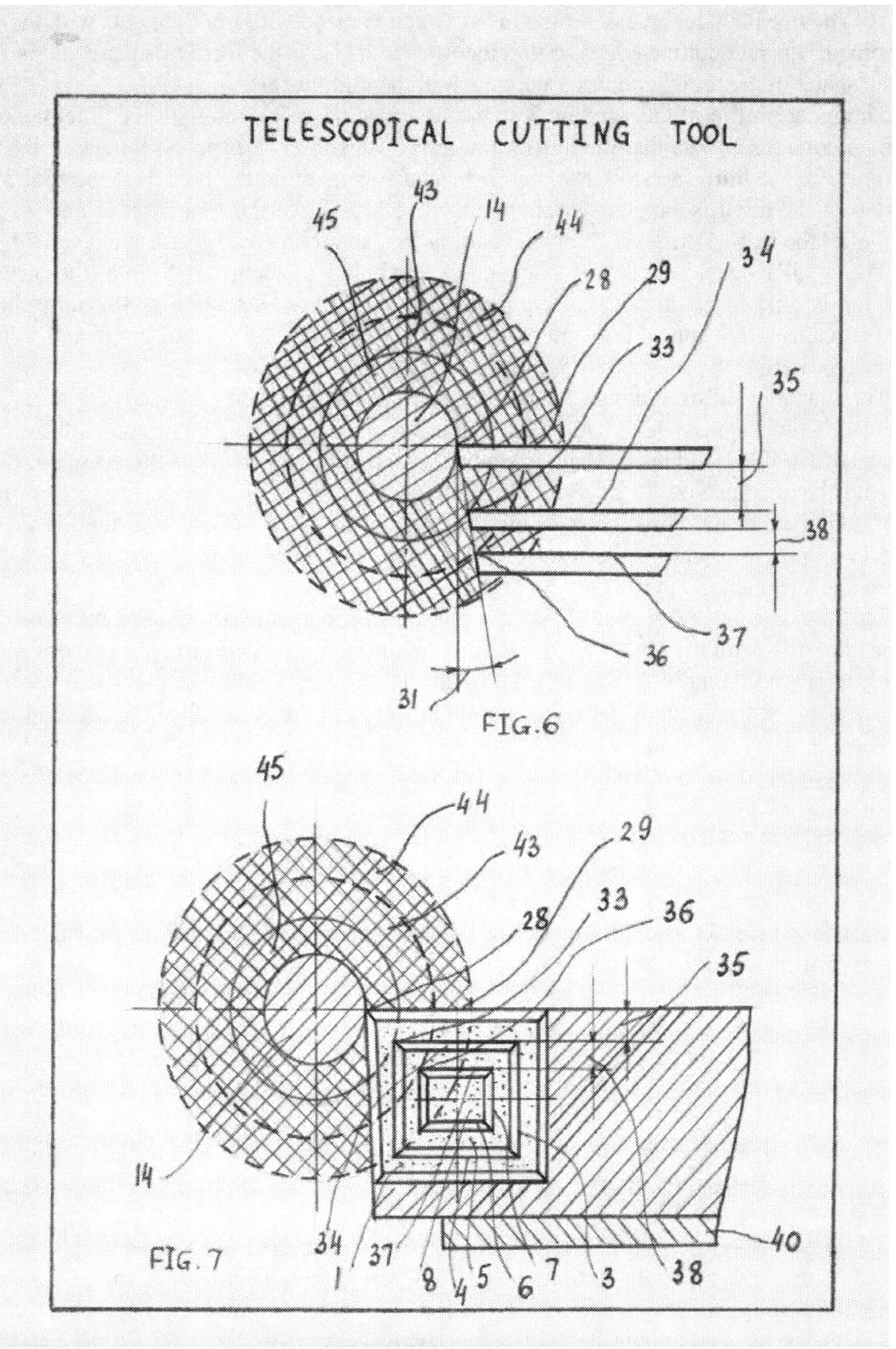

TELESCOPICAL CUTTING TOOL

FIG.6

FIG.7

This device includes an assembled insert which is made with multiple cuts with the combined, separate cutting edges so that the first insert 1 has the form of square and is fixed on the shank of the tool 2 and has a window 3 in the middle part.

The inner profile of the window 3 of the first insert 1 coincides with the outer contour 4 of the second insert 5 so that the inner contour 6 of this insert 5 is proportionate to the outer contour 7 of the third insert 8. The stepped canal 9 is made in the holder of the tool 2. The axis 10 of the canal is situated symmetrical to the axis 11 of the first insert 1 and arranged parallel to the axis of the feed 12 for the cutting tool and to the axis 13 for the detail 14.

The smaller size 15 of the cross-section of this hole is made in the form of the outer size 4 of the second insert 5, and the bigger size 16 of the cross-section at the same hole is proportionate to the movable element 17, which is in view of the hollow piston with the clamp 18 where the second insert 5 is fixed.

The movable hollow element 17 for fixing of the second insert 5 also has the non-hollow hole 19 of the stepped view. This axis of the movable hollow element comes through the axis 11 of the first fixed insert 1 and is situated parallel to the axis 13 of the detail 14.

And the smaller size 20 of the cross-section of the hole 19 is made in the form of the outer contour 7 of the third insert 8, and the bigger size 21 of the cross-section at the same hole 19 is proportionate to the movable elements 22 in view of the other piston with the clamp 23, on which the third insert 8 is fixed.

The movable hollow element 17 in view of the stepped piston for the second insert 5 has, at the clamp 18 with the bigger cross-section, the through hole 24 of the lower cross-section, which is situated axially to the axis 11 of the first insert 1 and has the space 19 of this hollow piston with the main space 9 of the first insert 1 in the holder 2 of the cutting tool.

Also, to the side of smaller size 20 cross-section on the movable hollow element 17 there are radial situated holes 25 with small holes that connects the inner space 19 of this piston 17 with the outer contour 4 of the cutting second insert 5.

So, the movable continuous piston 22 that fixes the third insert 8 to one side has a non-hollow hole, small diameter hole 26, which is situated axially, and the other side has a non-hollow, small diameter hole 27, that is placed radial to this piston 22, and in total, both these spaces of the holes 26 and 27 are connected with the inner space 19of the movable hollow piston 17 of the second cutting insert 5.

The peculiarities of the telescopical cutting tool consists in that the peak 28 and general cutting edge 29 of the first insert 1 are situated near the axis 13 of the cutting work-piece 14. The first cutting insert 1 has the front surface, which is made under a definite angle 30, and the back general surface is also under a definite angle 31. Also, the bevel 32 is done on the insert 1 and the holder of the tool 2.

The peak 33 and general cutting edge 34 of the second insert 5 are displaced downward, accordingly, with the first insert 1 on the definite value 35 depending on the cross-section of the insert 1 during the cutting process.

The peak 36 and general cutting edge 37 are displaced downward, accordingly, with the second cutting insert 5 on the definite value 38 depending on the cross-section of the insert 5 during the cutting process.

*In the process of cutting the work-piece with the increased allowance, the application of the new Telescopical cutting tool takes place in the following steps:

the work-piece 14, having total allowance 39, fastens to the chuck (not shown) and then installs the holder 2 of the turning tool in the tool-holder 40 and firmly fastens so that the peak 28 of the first insert 1 coincides with the axis 13 of the work-piece 14.

For improving the stiffness of the system and the telescopical cutting tool, it is necessary that the axis 11 of the movable hollow element 17 and solid element 22 –on which are fixed the first 1, second 5 and the third 8 inserts –be situated axially and parallel to the axis 13, to the work-piece 14, and also to the feed of the cutting tool. Then connection point 41 gives the feed lubrication 42 into the inner area 9 for moving out axially in the direction of the elements 17, 22, on which are fixed the inserts 5 and 8.

This forced lubrication 42 comes in the area 9 of the holder 2 cutting tool and later goes through the hole 24 of the movable element 17 to the area 19 and goes further through radial holes 25 to the cutting area for the turning allowance 43 by using the second cutting insert 5.

Through the axial hole 26 and the radial hole 27, this forced lubrication 42 comes to the working area, turning the allowance 44 by the third insert 8. So, the first insert 1 cut from the work-piece 14 is the allowance 45, and the second insert 5 cut from the work-piece 14 is the allowance 43, and the third insert 8 cut from the work-piece 14 is the allowance 44.

In view of the fact that the forced lubrication 42 comes under high pressure through the small diameter holes 27 and 25, the stiffness of system of movable elements 17 and 22 –on which are fixed the cutting inserts 5 and 8 –is expected.

At the cutting process, the lateral surface of the insert 1 arranges perpendicularly to the axis 13 of the cutting work-piece 14, and the lateral surfaces of the inserts 5 and 8 between it displace on the definite value of the step, for instance with the permanent or variable value 46 and 47.

Effectiveness of the Telescopical cutting tool is shown below.

EFFECTIVENESS OF THE TOOL

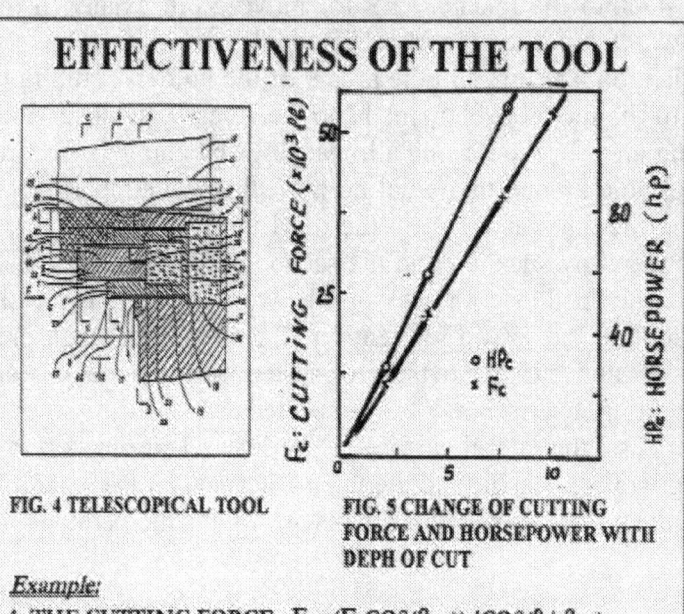

FIG. 4 TELESCOPICAL TOOL

FIG. 5 CHANGE OF CUTTING
FORCE AND HORSEPOWER WITH
DEPH OF CUT

Example:

1. THE CUTTING FORCE $F_C = (F_S COS(\beta-\alpha))/COS(\theta+\beta-\alpha)$
   F_S=SHEARING FORCE, $F_S = S_S*A_0*COSEC\ \theta$;
   α=RAKE ANGEL; θ=SHEAR ANGEL; β=FRICTION ANGEL,
   tan $\beta=\mu$; S_S=SHEAR STRENGTH OF MATERIAL;
   A_0=CROSS-SECTIONAL ARE OF CHIP;
   $A_0 = t*b$; t=DEPTH OF CUT; b=WIDTH OF CUT;

2. THE HORSEPOWER $HP_c = (F_C*S)/33000$;
   S=CUTTING SPEED, ft/min.
   MATERIAL 1045 STEEL, S_S=90000 psi; b=0.25 in; θ=43°,
   cosec 43°=1.466; tan $\beta=\mu$=0.96; β=43.8°; α=8°; S=90ft/min;

t=0.075 in;	F_C=10418lb(4735kg);	HP_c=28hp(21kw);
t=0.157 in;	F_C=21809lb(9913kg);	HP_c=59hp(44kw);
t=0.393 in;	F_C=54592lb(24814kg);	HP_c=147hp(109kw);

3. Rozenblat's combined die

The varieties of multi-operational dies relate to Rozenblat's combined die is shown in Figures 1,2,3 and 4.

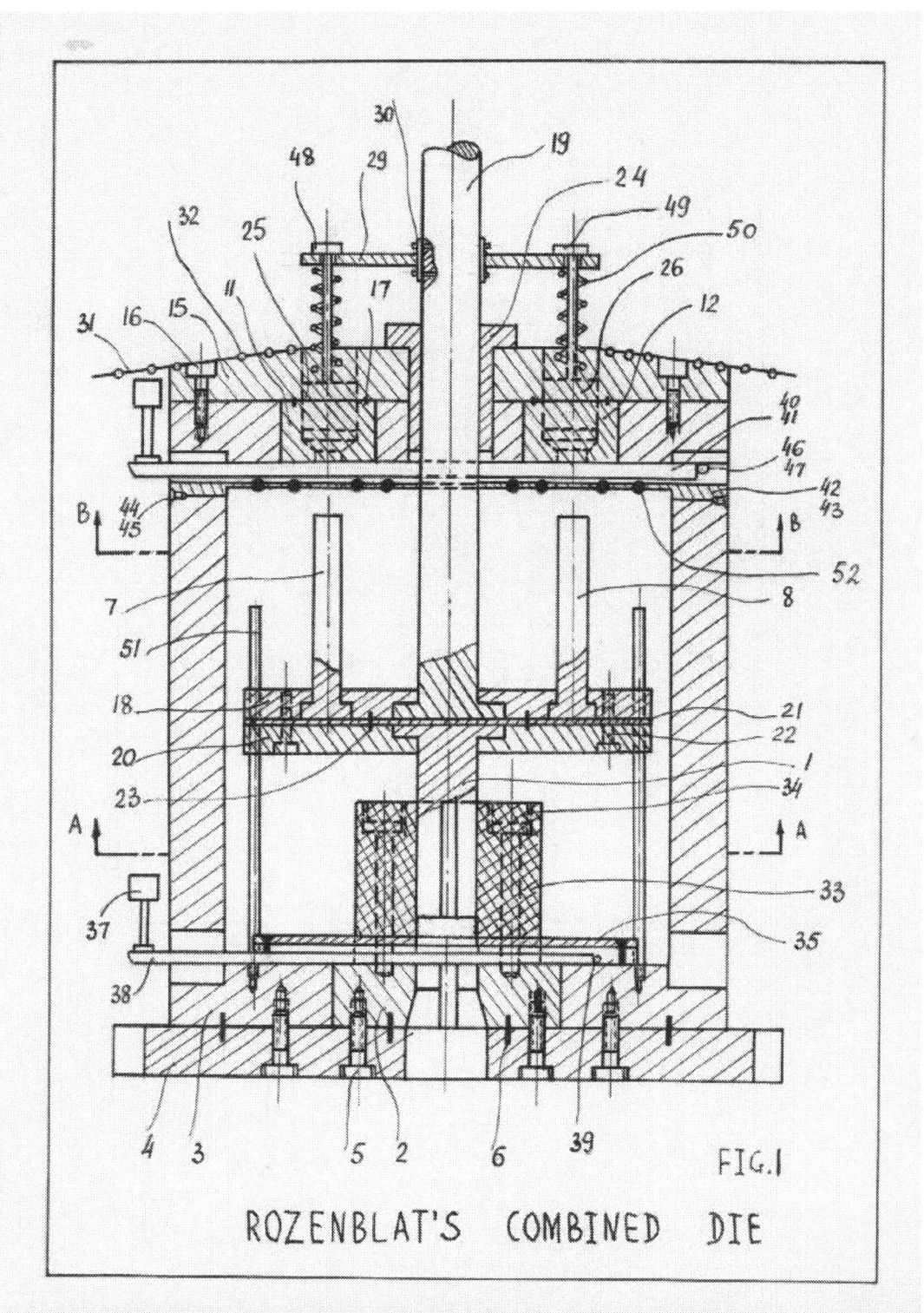

FIG.1

ROZENBLAT'S COMBINED DIE

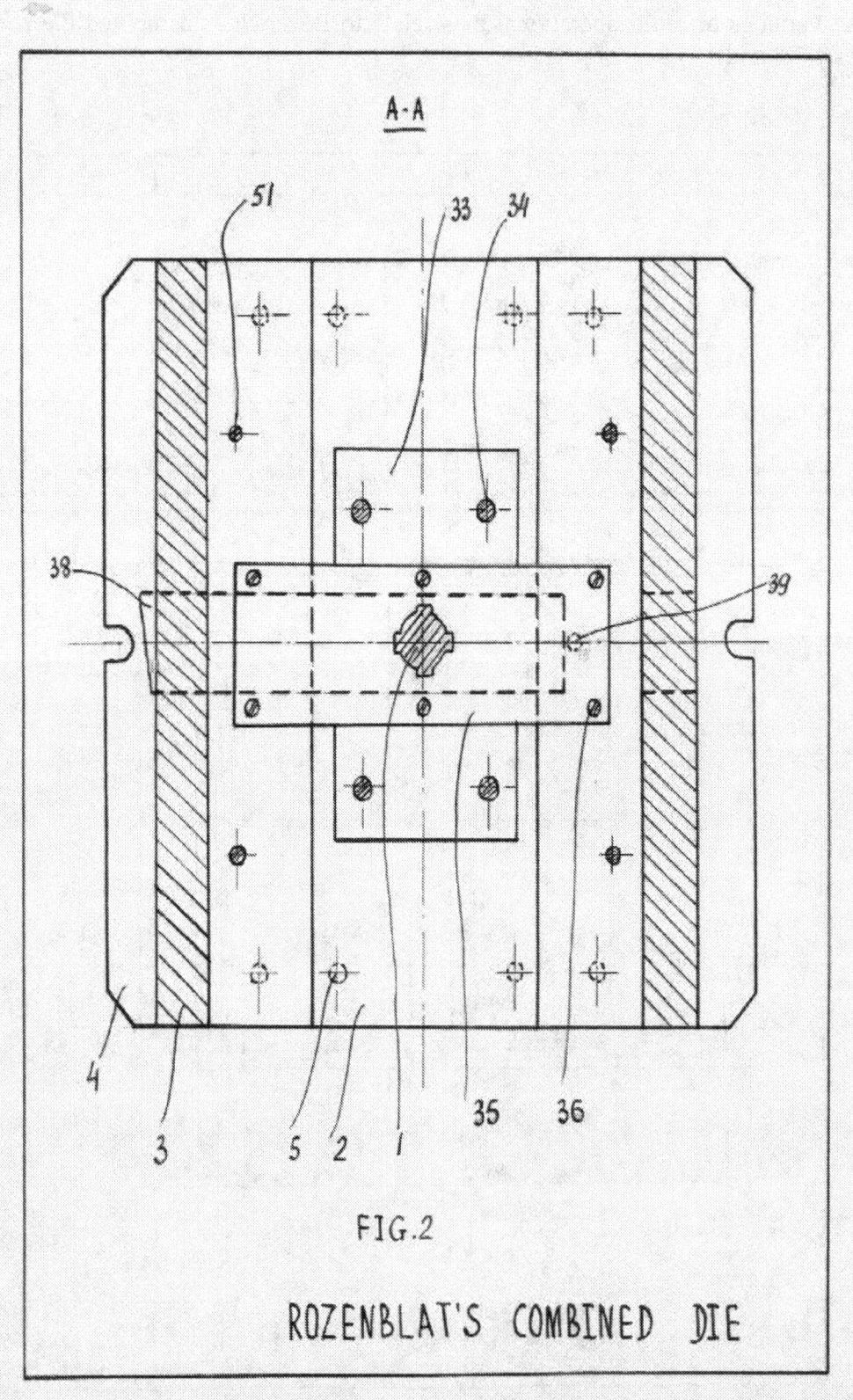

A-A

FIG.2

ROZENBLAT'S COMBINED DIE

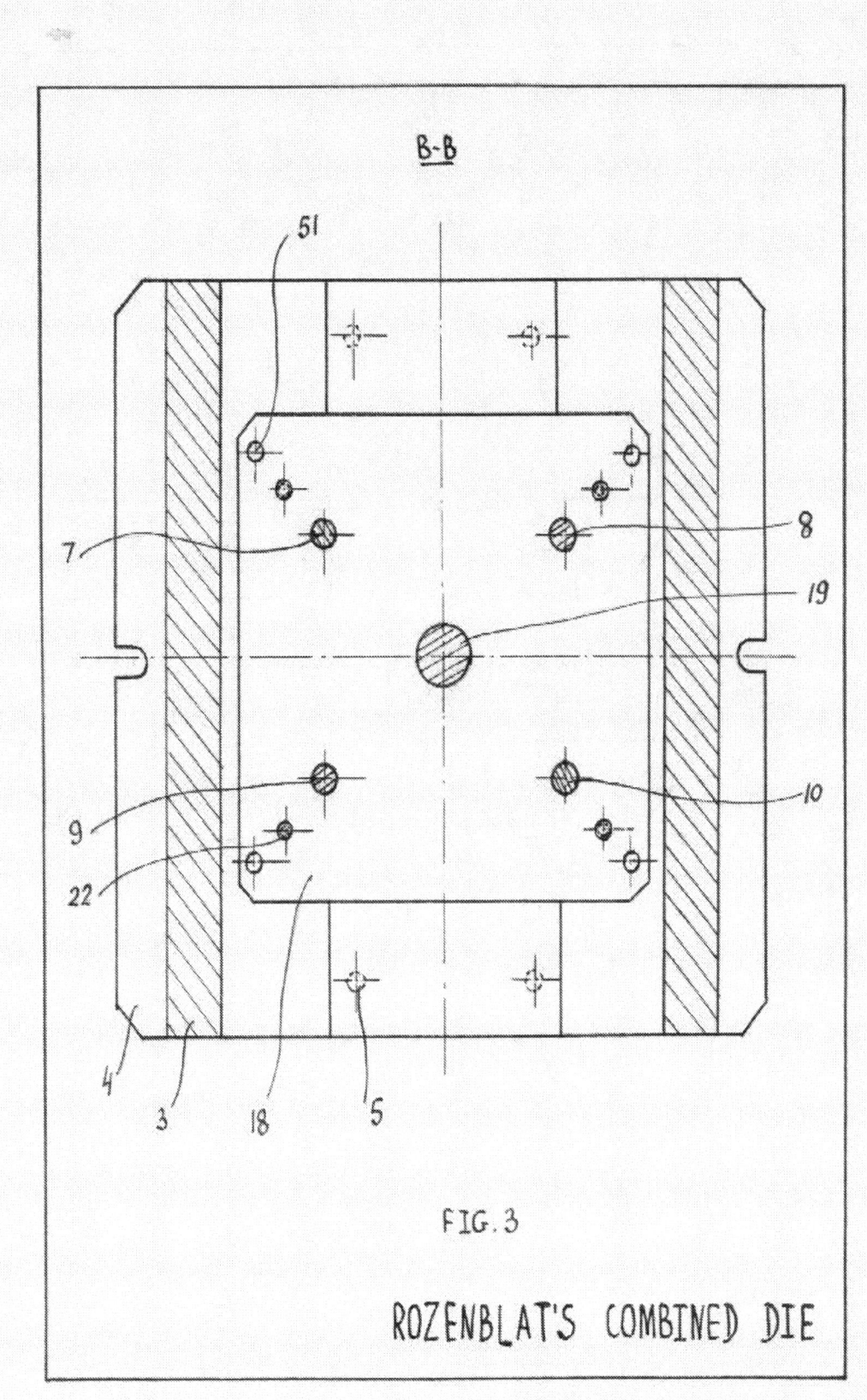

FIG. 3

ROZENBLAT'S COMBINED DIE

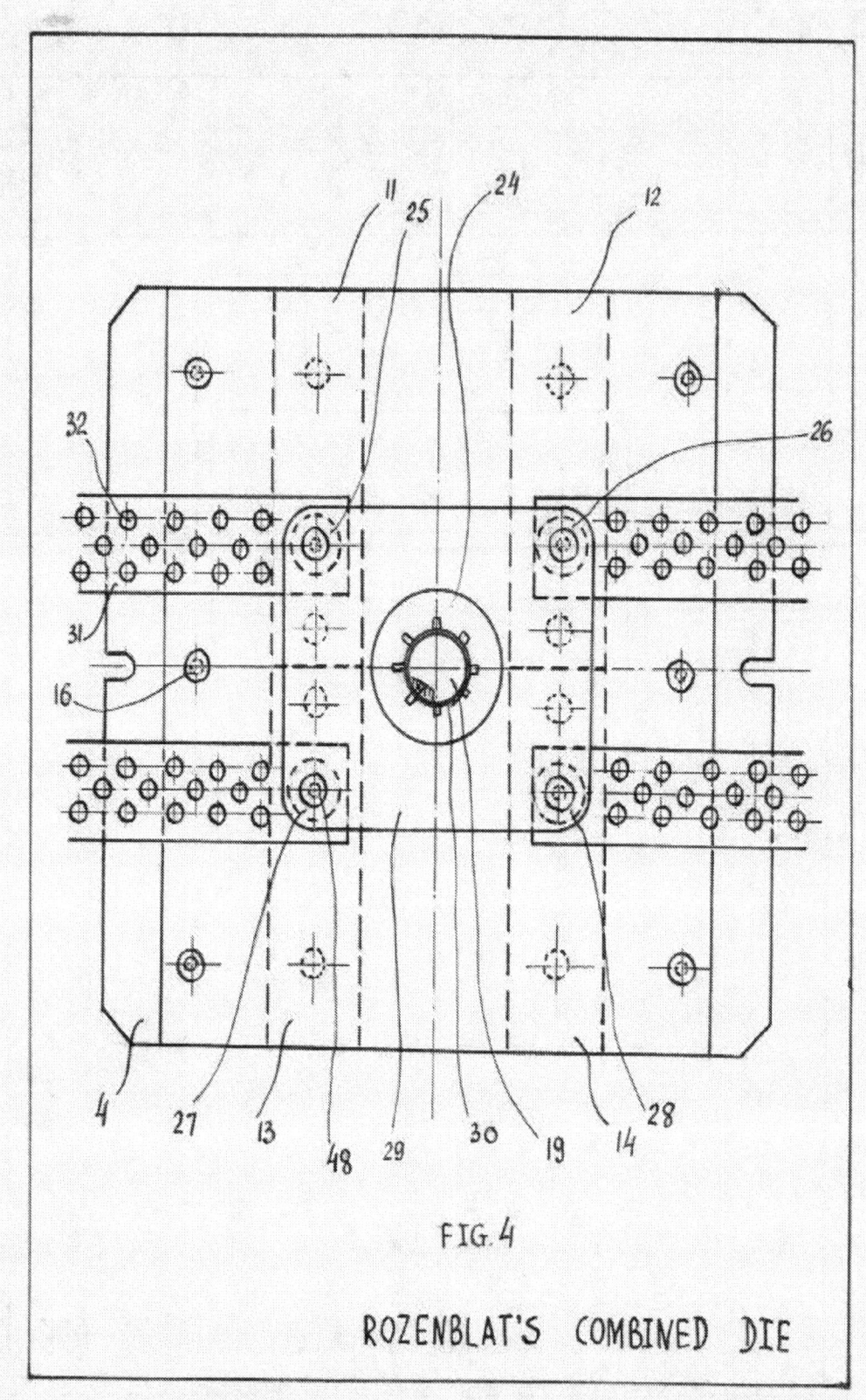

FIG. 4

ROZENBLAT'S COMBINED DIE

The combined die includes such main elements as a punch 1 and die 2 which are fixed in the body 3 and to the lower plate 4 by screws 5 and pins 6. In the upper part of the body 3, besides of main elements 1 and 2 the additional system is arranged in view of multiple pairs of punches 7 and 8,9 and 10 in mirror image to each other and also the other multiple pair of dies 11,12 and 13,14 conforming to the punches 7,8 and 9,10.

The dies 11,12 and 13,14 fasten to the upper plate 15 and to the body 3 with screws 16 and pins 17. The punches 7,8 and 9,10 fasten to the punch holder 18which has a hole leading to the axis of the press and the assembled stamp for placing of the shank 19.

The punches 7,8,9 and 10 are placed and fixed firmly to the plate 20 with screws 22 and pins 23. The upper plate 15 of the axis has a hole and fastens the guide bushing 24which interacts with the shank 19. Also in the upper plate 15 in the outer zone of the dies 11,12 and 13,14 the system of electromagnets 25,26 and 27,28 is placed and joined in a common frame 29 by means of a clamp 30 on the shank 19.

In the outer zone of the sheet bar, the chamfers 31 with arranged bearings 32 are made from dies 11,12 and 13,14 on the upper plate 15. With the aim of adjustment and cushioning, the assembled stamp has a buffer system 33 such as rubber which is fixed on the die 2 or on the punch holder 20 by means of screws 34.

- The process of adjustment and the working process of stamping in the combined die of A.I.Rozenblat works in the following way:

On the guide strip 35, which is fixed by means of a screw 36, the sheet bar 38 moves by means of manipulator 37 against the stop 39 to the die 2 and body 3. The movement of the punch 1 in the lower position (the working process of stamping) creates the cutting-off process on the sheet bar 38 upon the next removal of the work-piece from the cutting area.

At the movement of the press to the upper position, the punches 7,8 and 9,10 accordingly contact with the arranged sheet bars 40 and 41 which are held to the strips 42 and 43 by means of screws 44 and 45. Special stops 46 and 47 are made for orientation and cut-off sheet bars 40 and 41.

In view of the fact that the punches 7,8 and 9,10 drive the working process, the made sheet bars that are made and come in the holes of the dies 11,12 and 13,14 are removed later from by means of electromagnets 25,26 and 27,28 which are fixed by means of a frame 48 and 49 that has a reverse spring 50.

The construction of the die also has special guide columns 51 for safety. With the aim of keeping the sheet bars 40 and 41 at the working process the special rollers 52 on the guide strips 42 and 43 have been made.

4. Break-chips insert

The other variety of cutting tool in the question of improvement of break chips is the Break-chips insert which is shown in Figures 1,2,3,4,5,6,7,8,9,10,11,12 and 13.

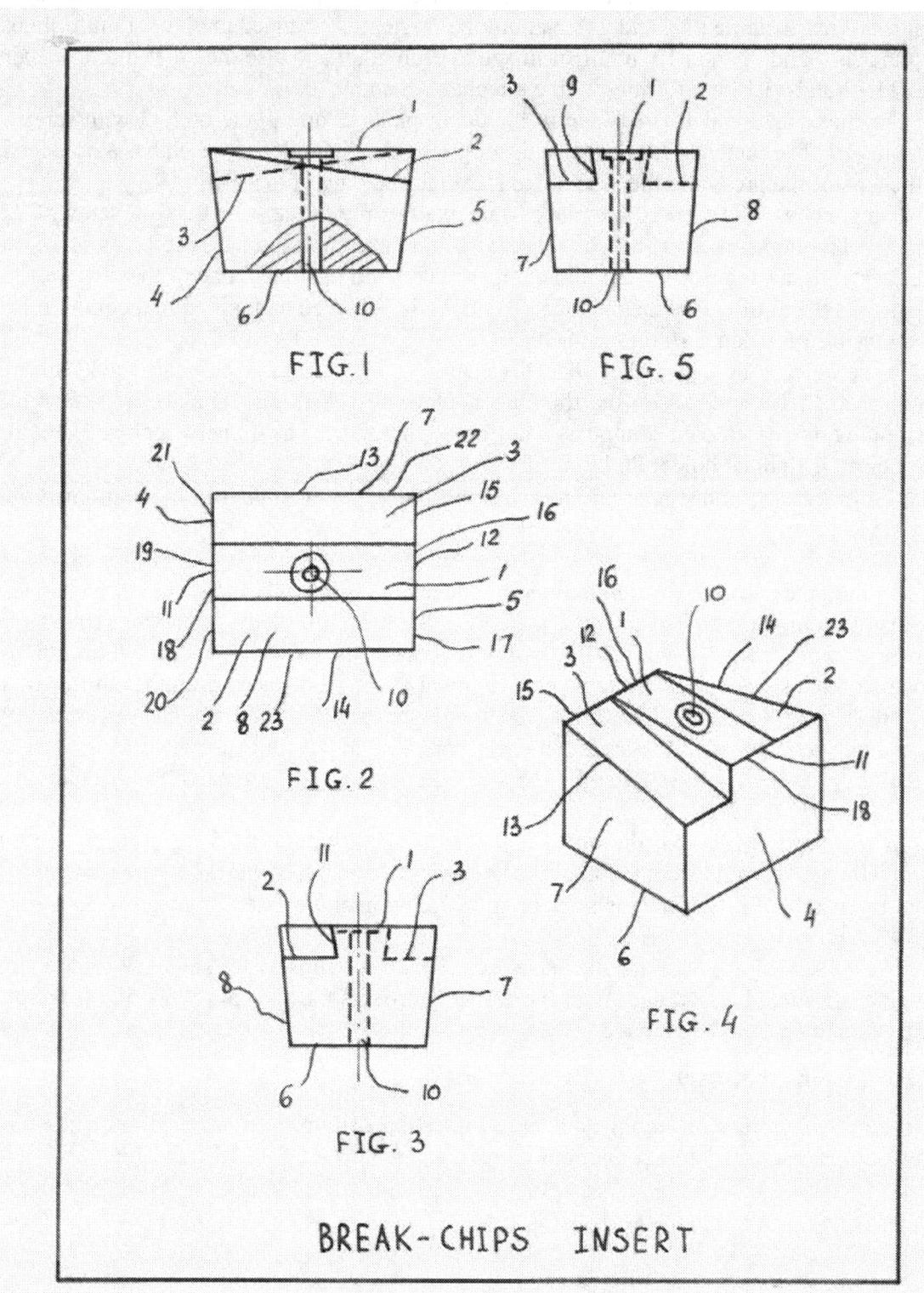

FIG.1

FIG. 5

FIG.2

FIG. 3

FIG. 4

BREAK-CHIPS INSERT

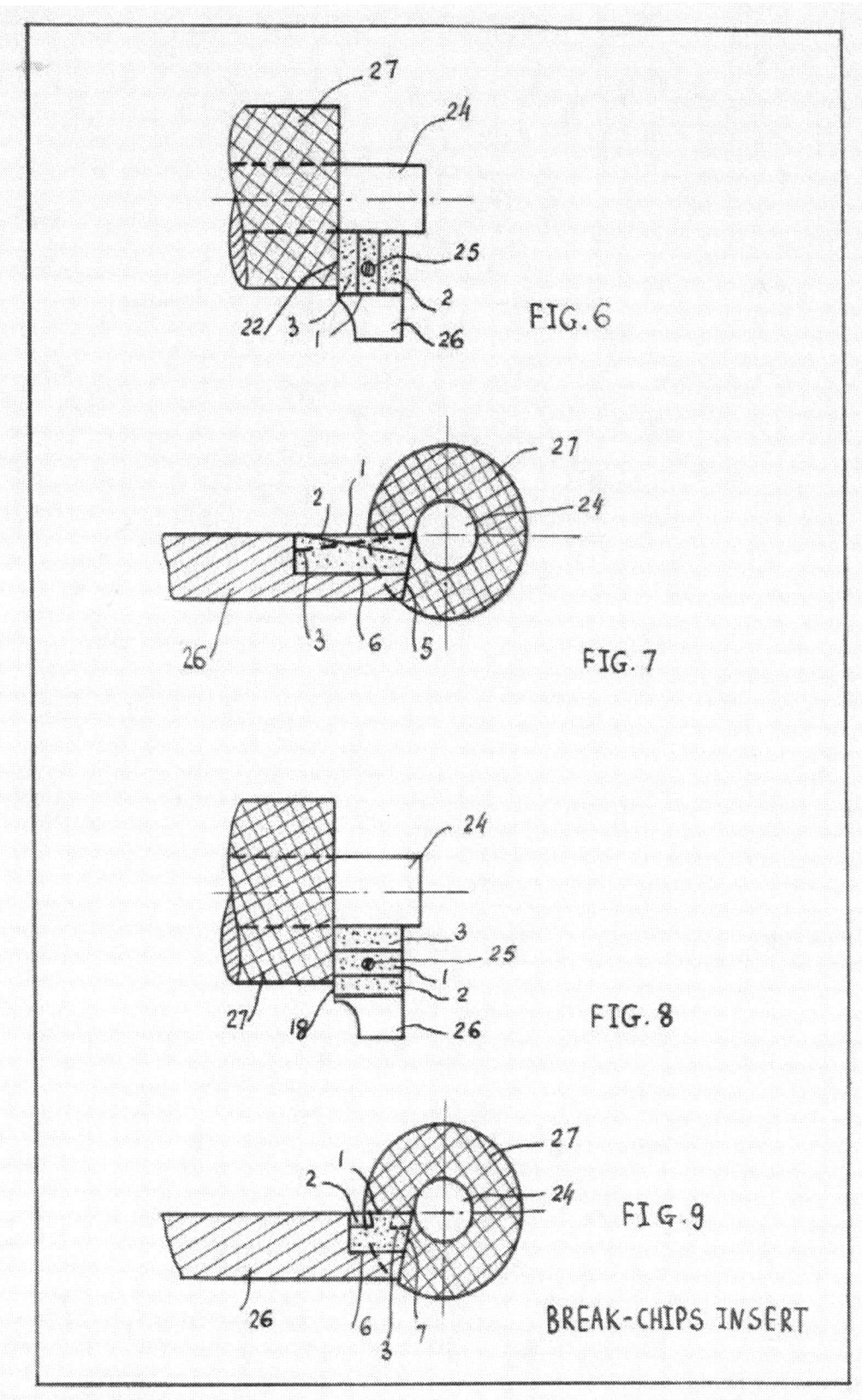

FIG.6

FIG.7

FIG.8

FIG.9

BREAK-CHIPS INSERT

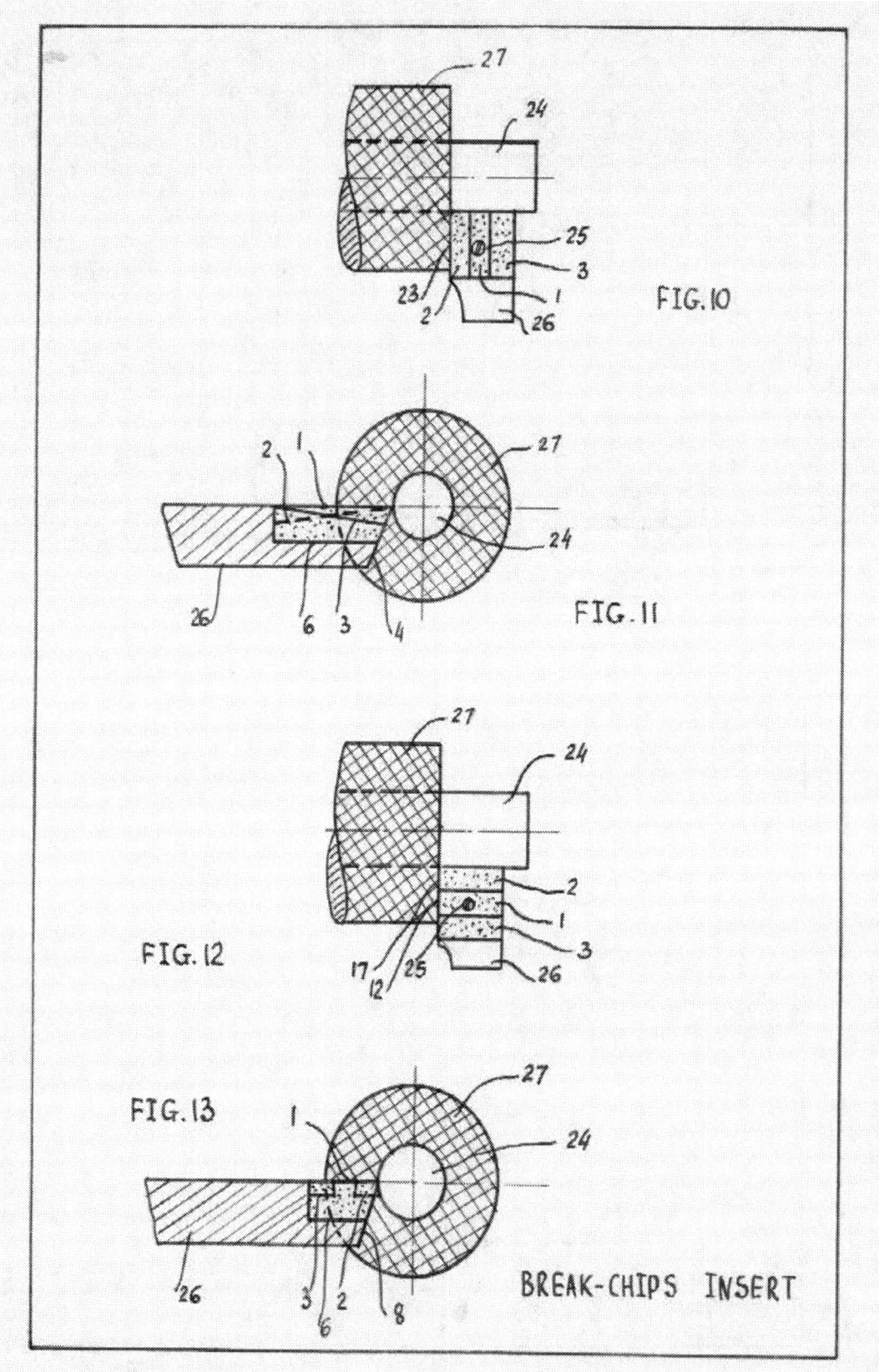

FIG.10

FIG.11

FIG.12

FIG.13

BREAK-CHIPS INSERT

The insert includes the first separate front surface 1 which contacts the second separate front surface 2 and the third separate front surface 3. The break—chips insert has the profile plane 4 and 5 which contacts the above-named surfaces 1,2 and 3, and also the plane 6.

Besides this, the insert has the profile planes 7 and 8 which close the contour of the insert to the perimeter. The third and the second planes 3 and 2 that contact the first front plane 1 are made with the back angle 9.

To fix the break-chips insert, having three planes to the holder of the cutting tool provides the hole 10. The first front plane 1 of the insert is set with a zero value of the front angle, the second front plane 2 is set with the positive front angle, and the third front plane 3 is made with the negative front angle.

As we see, three separate surfaces 1,2 and 3 are joined between themselves so that each separate plane can be made with any variable front angle as shown in Table 1.

TABLE 1 Designing of new Break-chips insert with the variable front angles

Type	Number of plane	Value the front angle, degree		
		Zero	Positive	Negative
I	1	x		
	2		x	
	3			x
II	1		x	
	2	x		
	3			x
III	1			x
	2	x		
	3		x	
IV	1	x		
	2			x
	3		x	
V	1		x	
	2			x
	3	x		
VI	1			x
	2		x	
	3	x		

• In the cutting insert, three separate planes 1,2 and 3, along with the profile surfaces 4,5 and 7,8 form the four main cutting edges: two of them 11 and 12 are the broken lines and the other two 13 and 14 are straight lines.

In this instance, the first main cutting edge 12 is in view of the broken lines and forms the three auxiliary cutting edges 15,16 and 17 which are connected in during the total cutting peak, and the third cutting edge 17 is displaced relative to them at some front (positive) angle.

Also, the second main cutting edge 18 comes into view of the broken lines and forms three auxiliary cutting edges so that two of them 19 and 20 are displaced in mirror image and arranged oppositely and connected between themselves during the total cutting peak, and the third 21 is displaced relative to both cutting edges at a front (negative) angle.

The third main cutting edge 22 is drawn as a straight line and is displaced relative to the total cutting peak on a negative front angle.

And the fourth main cutting edge 23 also is drawn as a straight line and displaced relative to the total cutting peak on a positive front angle.

- The method of cutting a work-piece with the use of a Break-chips insert consists of the following steps:

The work-piece 24 fastens firmly in the chuck. And the break-chips insert with the cutting surfaces 1,2 and 3 also fastens on the axis 25 in the turning tool holder 26 so that the third main cutting edge 22, with the negative front angle has contact with the cutting total allowance.

As a result of wear, the third main cutting edge 22 uses the next cutting edge-the second main cutting edge 18, which is situated relative to the above-named total allowance 27 so that this total cutting edge is in view of the broken lines that separately have three planes 1,2 and 3—in case of the need to use the fourth main cutting edge 23 for cutting total allowance 27 use, for instance the front cutting plane with the positive front angle 2.

And finally at the cutting of the work-piece 24 there is also the possibility of using the first main cutting edge 12, which is situated relative to the total allowance 27 so that this front plane is in view of the total broken lines, and the auxiliary cutting edge also is made with a positive front angle.

The advantages of the Break-chips insert are shown below.

TOOLS FOR BREAK CHIPS

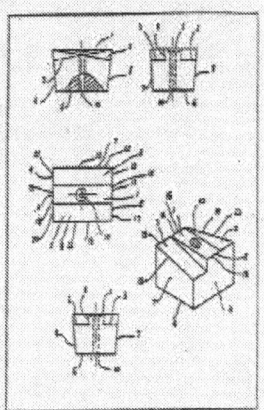

FIG. 6 BREAK CHIPS INSERT

ADVANTAGES:

1. THE BREAK CHIPS PROCESS IMPROVES

2. THE TECHNOLOGICAL POSIBILITY OF TOOLS EXPAND.

3. SAFETY OF MACHINIST IMPROVES PARTI-CULARLY FOR TURNING STAINLESS AND HIGH-RESISTANCE STEELS.

4. AUTOMATIZATION OF CUTTING PROCESS IMPROVES.

5. USING OF PNEUMATIC TRANSPORTATION FOR REMOVAL OF CHIP FROM ZONE OF CUTTING

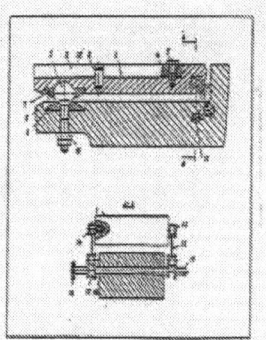

FIG. 7 TOOL WITH VARIABLE ANGLES

5. Multiple carbide insert and methods of its designing

One variety of designing of the new cutting tool for turning the work-piece with the large allowance is a Multiple carbide insert, which is shown in Figures 1,2,3,4 and 5.

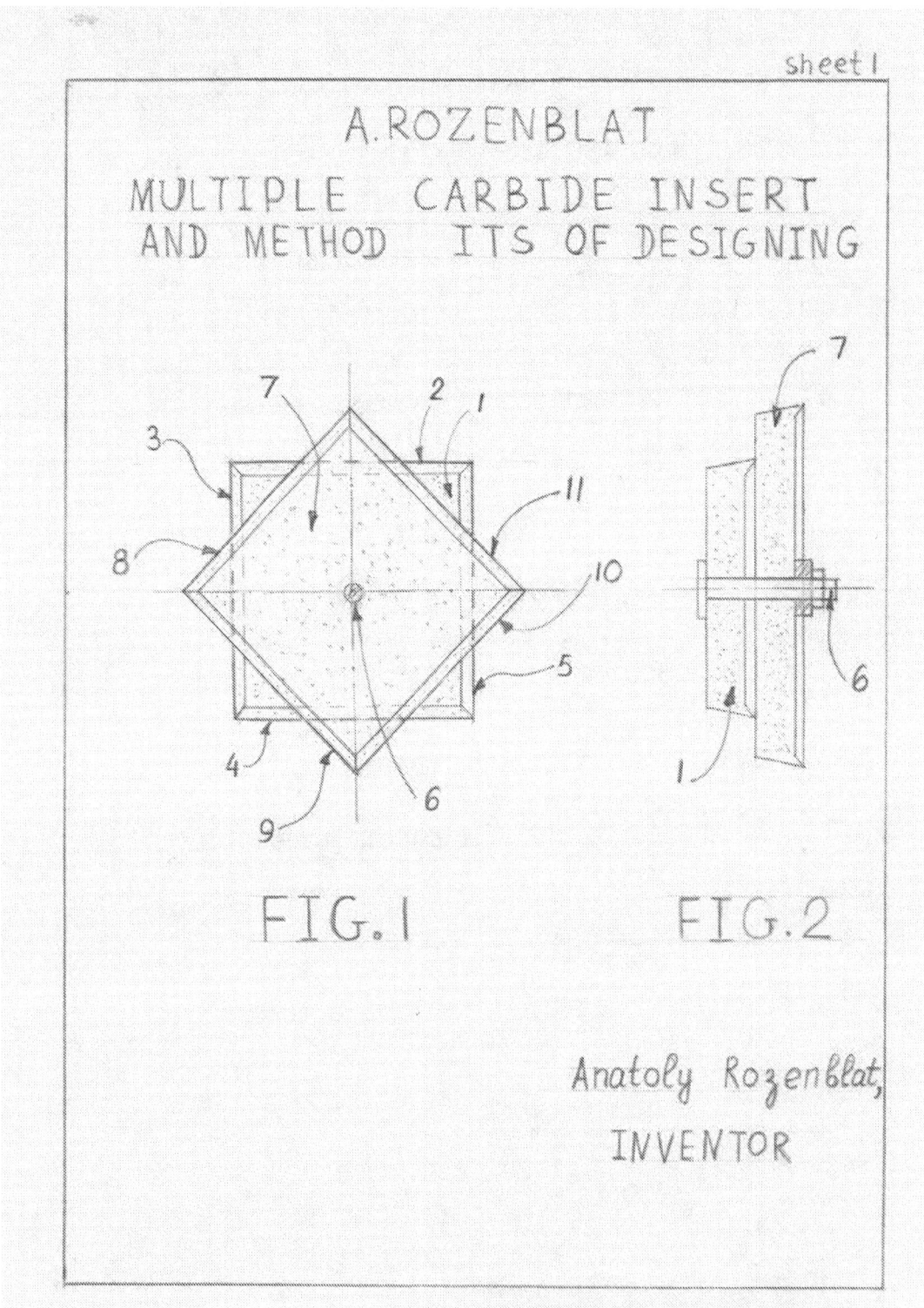

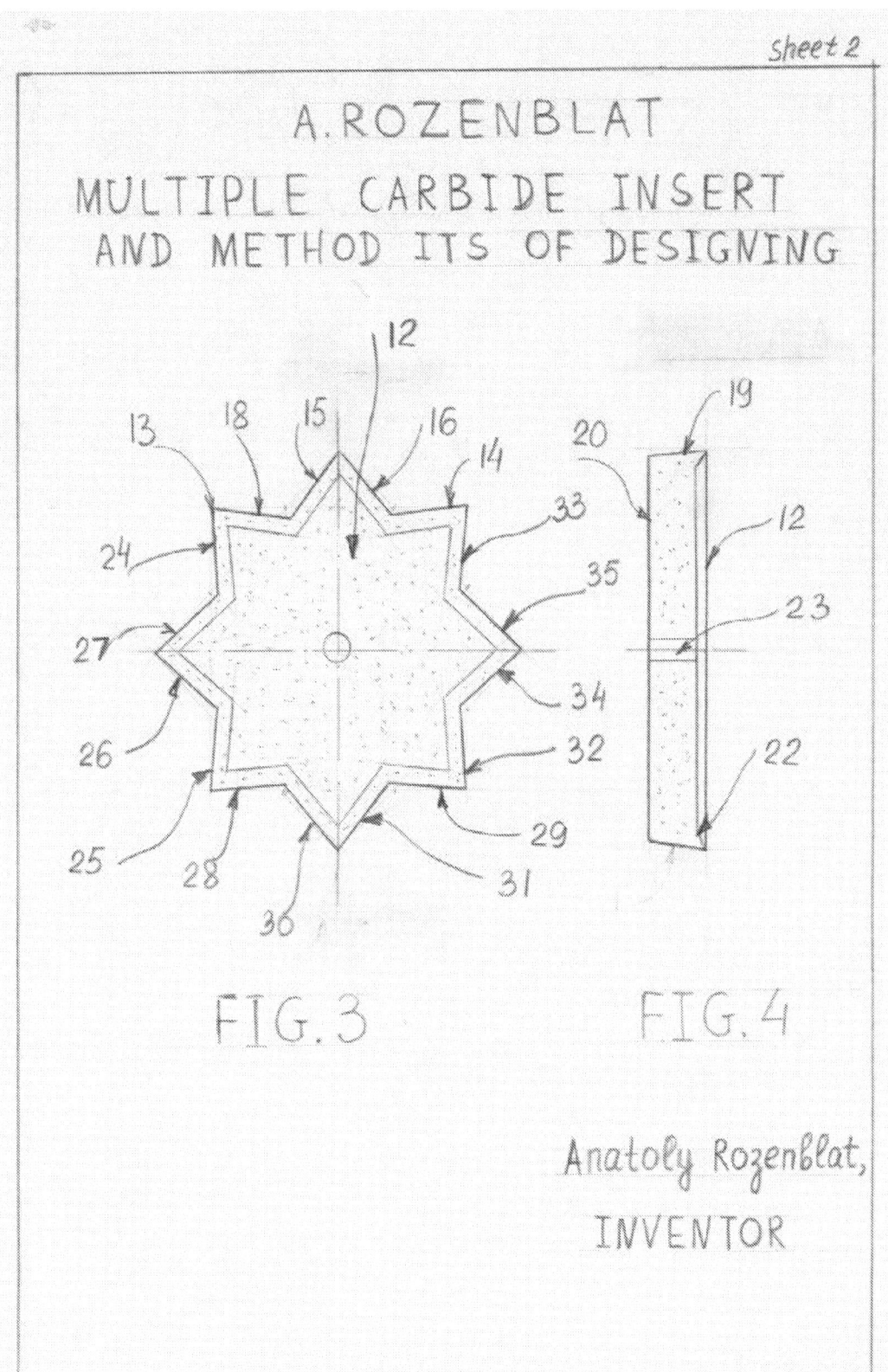

A.ROZENBLAT

MULTIPLE CARBIDE INSERT
AND METHOD ITS OF DESIGNING

FIG.3   FIG.4

Anatoly Rozenblat,
INVENTOR

Anatoly Rozenblat

A.ROZENBLAT

MULTIPLE CARBIDE INSERT
AND METHOD ITS OF DESIGNING

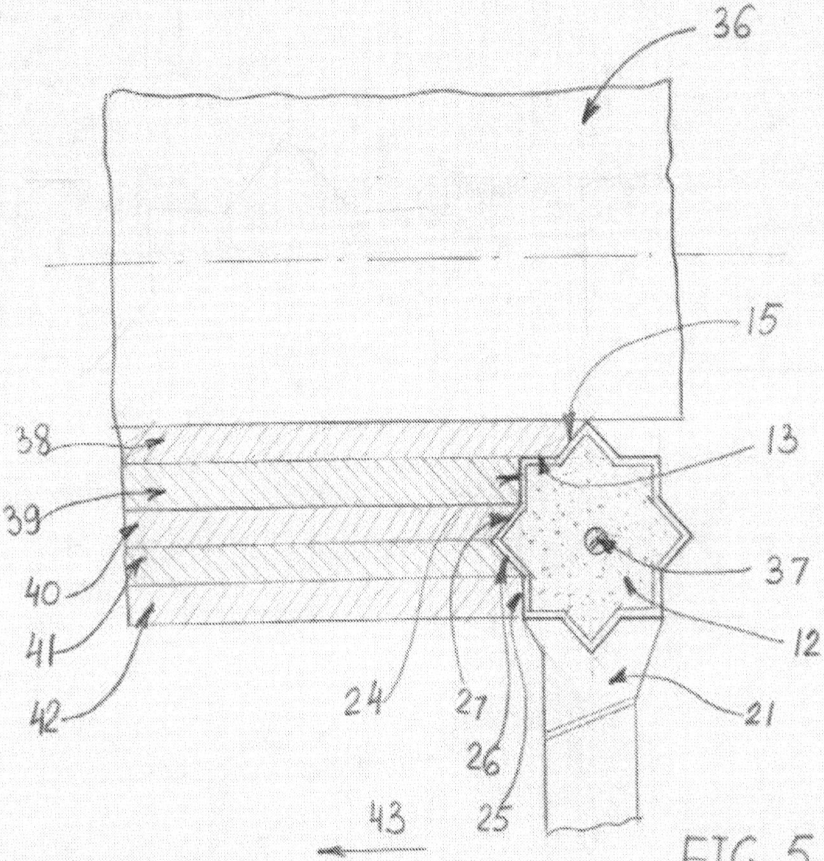

FIG.5

Anatoly Rozenblat,
INVENTOR

72

The method of forming the multiple carbide insert consists of the following steps:

The basic plate 1 with the general cutting edges 2,3,4 and 5 on the perimeter (in view of the square) puts on the axe 6 and fixes itself. Then the auxiliary plate 7, with additional cutting edges 8,9,10 and 11 on the perimeter [in view of the square estimated with the first base plate 1 also puts on the axe 6, and the auxiliary plate 7 turns relative to the base plate 1 on the 45-degree angle and fixes itself, and projects the new contour of the many-sided insert on the frontal surface by methods of photographing or designing.

The author also admits that in such a method of forming the multiple carbide insert in view of the base 1 and auxiliary plates 2 the use of plates of different forms and sizes are estimated with the machining detail of cutting. In this form of the new many-sided profile cutting tool, the method is created of putting two or more compounds of element plates, differing in form and size, to use for projecting of the multiple carbide insert.

So, in this method the new many-sided profile of the multiple carbide insert 12 forms, having four equal cutting edges joined one with the other in view of general cutting edges 13 and 14, and also of additional cutting edges 15 and 16, coming together with the general cutting edges 13 and 14 at angles of 45-degrees.

In summary, the new cutting edges in view of the other edges 13,14,15 and 16 are fulfilled in view of the line so that the cutting edges 13 and 14 have the auxiliary back angle 18 within the limit of one to three degrees for removal of chip generated in the process of machining detail by cutting beside the base of the back angle 19.

The support surface 20 of the multiple carbide insert 12 adjoins to the tool holder 21 in the process of the installation of the tool and the machining of detail by cutting. The frontal surface of the multiple carbide insert 12 has small front angles 22 on each of the cutting edges 13,14,15 and 16.

Fox fixing of the multiple carbide insert 12 to the tool holder 21 the base hole 23 is used. Each general cutting edge consists of two base cutting edges 13,14 and two auxiliary cutting edges 15 and 16, but in sum total they join on the perimeter many-sided insert with symmetrically equally situated new cutting edges relative to the base hole 23.

So, the first new total cutting contour has two general cutting edges 13 and 14, and also additional cutting edges 15 and 16. The second new total cutting contour has two general cutting edges 24,25 and also additional cutting edges 26,27;the third new total cutting contour has two general cutting edges 28, 29 and also additional cutting edges 30,31;the fourth new total cutting contour has two general cutting edges 32,33 and also additional cutting edges 34 and 35. In the process of machining detail 36 by cutting, the axe 23 is fixed on the multiple carbide insert 12 with the help of a special installation 37 on the tool holder 21 that participates in the cut of the first allowance 38, the additional cutting edge 15 with 45-degree angles, and the general cutting edge 13.

So, this improves and reduces the temperature heating of the detail 36. In the process of cutting the second allowance 39, the general cutting edges 24 and 13 participate. In the process of cutting the third allowance 40, the additional cutting edge 27 participates. In the process of cutting the fourth allowance 41 the additional cutting edge 26 participates. In the process of cutting the fifth allowance 42 the general cutting edge 25 participates.

So, we see that in the process of machining detail 36 in a one-way cut (to the feed 43 0 five allowances take place from the detail suggested on the multiple carbide insert 12 fixed on the tool, and this improves the process of machining and reduces the temperature and deformation of detail and accordingly increases the precision of the detail in whole.

Description of advantages of the invention "MULTIPLE CARBIDE INSERT AND METHOD OF ITS DESIGNING"

EFFECTIVINESS:

In this device provides by means of rising the precision and quality of detail in process of cutting into account of decreasing the temperature deformation of detail.

NOVELTY AND ORIGINALITY:

Multiple carbide insert has the general cutting edges on perimeter so that each new general cutting edge is made in view of broken up line and consists from the four equal cutting edges which are joined one with another and in sum they united on perimeter of this multiple insert.

And besides the method of forming such multiple carbide insert supplies by means that primary one insert with the cutting edges on perimeter, for instance in view of square, puts on the axis of holder tool and then the other insert with the cutting edges also, for instance in view of square, puts stepped on the axe of the holder tool by that the first plate (insert) fixes and the second insert rotates relatively of the first insert under of any angle and fixes, and then projects the new contour of many-sided insert on the frontal surface, for instance by the methods of photographing or designing.

UTILITY AND ADVANTAGES:

Multiple carbide insert permit to cut from detail the big allowance because in the process of cutting by this tool participates many cutting edges and this all operation and process makes up in one way.

And besides the last cutting edges adjoining to the surface of detail situated under the angle, for instance of 45-degree and this all guarantees lower temperature of heating of detail and increase the precision of its in whole.

6. Use of arc plasmatron in pneumatic transportation

Pneumatic transport now widely is used for removing a variety of materials, including ferromagnetic particles such as iron or steel, produced as part of industrial manufacturing processes, usually turning and milling. In Figures 1, 2 and 3 is shown pneumatic installation and use of arc plasmotron at this system.

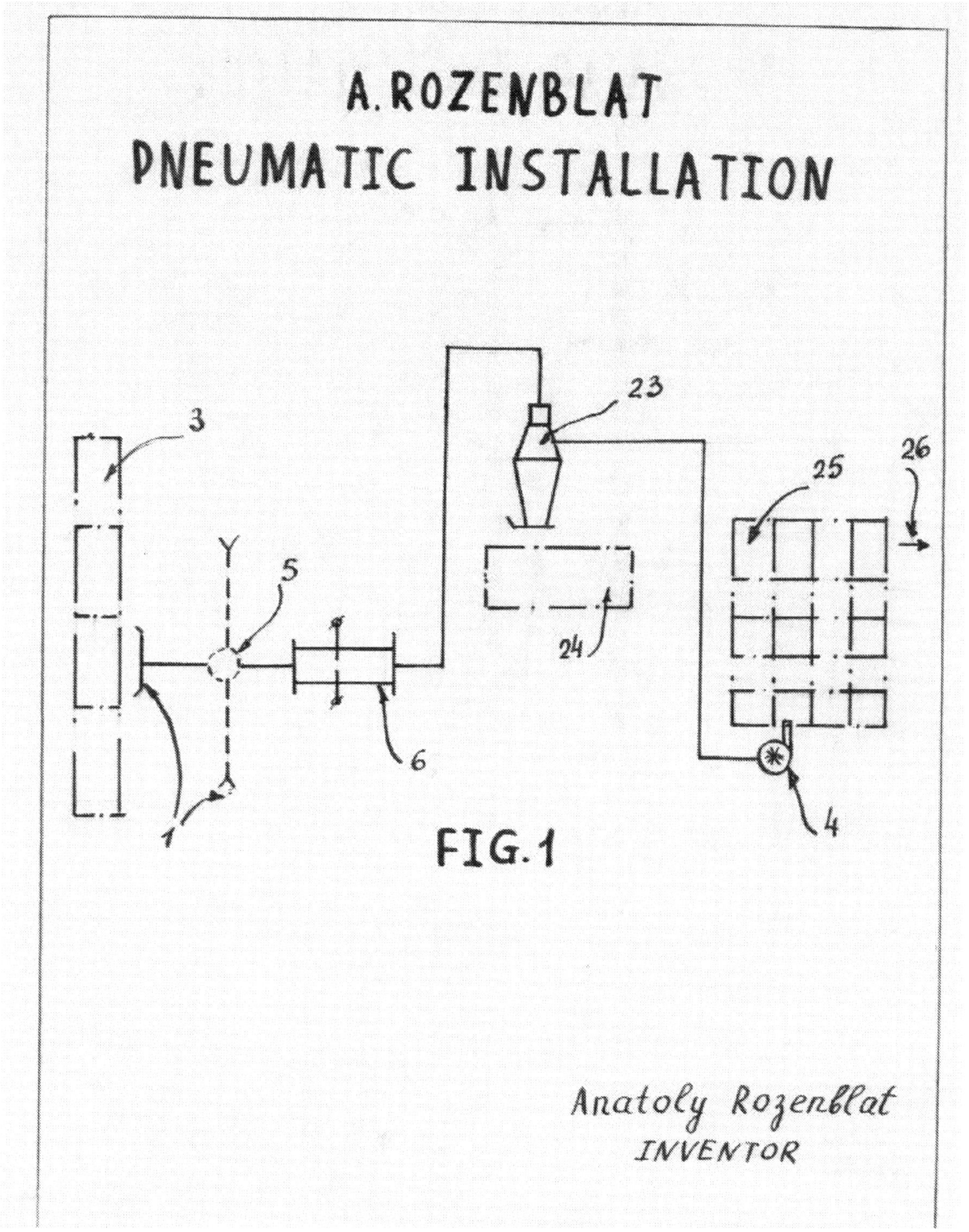

A.ROZENBLAT
PNEUMATIC INSTALLATION

FIG.1

Anatoly Rozenblat
INVENTOR

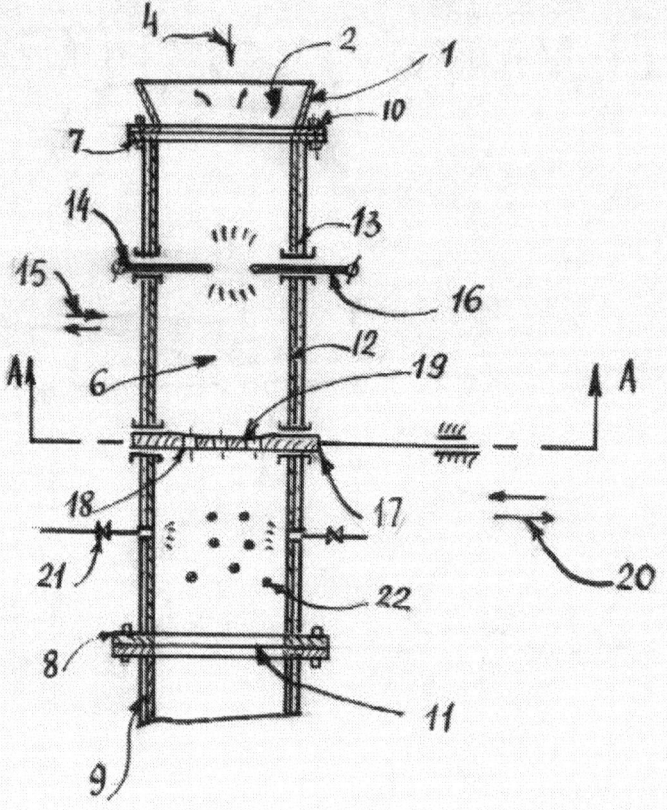

A.ROZENBLAT
PNEUMATIC INSTALLATION

FIG. 2

Anatoly Rozenblat
INVENTOR

A.ROZENBLAT
PNEUMATIC INSTALLATION
<u>A-A</u>

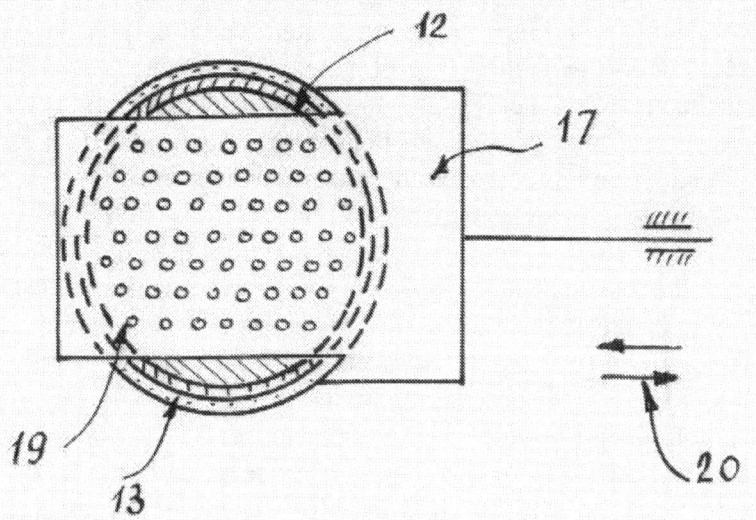

FIG.3

Anatoly Rozenblat

INVENTOR

Referring to the Figure 1 we see that the pneumatic system consists of the dust-chip arrangement 1, which removes the ferromagnetic chips 2 from the cutting zone of the group of fixed machine tools 3 by use of a vacuum of centrifugal ventilation 4. It is important to point out that machine tools 3 usually are joined in the automatic working line and have a total pneumatic collector 5 for removing of ferromagnetic material through the pneumatic system. However, such ferromagnetic particles have a very high abrasion index that causes continual wear and damage to the pipeline which transports it. Pneumatic transportation of such material considerably decreases the productivity of the process as a whole because the forces of friction for the movable material in the pipeline system are considerable.

The most rational way of removing these disadvantages is by transforming the shape of ferromagnetic material (chip) into a granulated (spherical) shape. This considerably changes the forces of fiction generated by rolling. So, the process of transforming the shape of the ferromagnetic material 2 necessitates the use of a special arc plasmatron 6.

As shown in Figure 2,3, the arc plasmatron 6 consists of an assembled cylinder having the pipe flanges 7 and 8 for upper and lower parts. By means of the flanges, this plasma-tron 6 fixes to the dust-chip arrangement 1 and pneumatic system 9 with screws 10.

The thermo-insulator packings11 are installed in the pipe flanges 7 and 8. The assembled cylinder of the plasmatron 6 is made up of the cartridge 12 and the thermo-insulator trimming 13. In the upper part of the plasmatron 6 two carbon rods are introduced. One carbon rod 14 is movable and commits the relapsing-progressive movement 15, and the other carbon rod 16 is unmovable. In the middle part of the plasmatron 6 the movable element is introduced in view of the plain matrix 17. The matrix has multiple holes 18 so that the frontal surface 19 of this matrix 17 is concave in the form of the cylindrical surface of the body 12 of the plasmatron 6 and also has the relapsing-progressive movement 20 as shown in Figure 3. In the lower part of the plasmatron 6, on the path of the transported material, the water or sprinkler system 21 for cooling of this material is installed. In the process of cutting material, the ferromagnetic chips 2 from the zone of the machine tool 3 come by means of the vacuum from the primary centrifugal ventilator 4 to the dust-chip arrangement 1 and moves to the pneumatic pipeline and plasmatron 6.

In the plasmatron 6, this material 2 melts under high temperatures as this area supports the impulse arc by means of two carbon rods 14 and 16. At this stage, the transported chips 2, now a liquid mass, moves further to the zone of the matrix 17, coming through multiple holes 18 [the presence of the concave surface 19 on the matrix 17 improves the gathering of melted material] and then this material reforms in the shape of these holes 18. In the lower part of this matrix 17 the water or sprinkler system 21 improves the receiving of the granules (spheres 22) from the transported material 2. The material, in the form of granules 22, moves further with the assistance of the vacuum of the centrifugal ventilator 4 to the cyclone 23 where this material 23 separates from air and settles in the capacity 24 (this operation is possible to do during stamping or other processes). The hot air from the pneumatic system (the plasmatron supports the very high temperature from the arc) comes through the ventilator 4 and can be used for any system 25 (heating of building, etc.). It must be noted that the process of removal ferromagnetic chips 2 from the cutting area of the machine tools 3 ensures the long life and safety of the pneumatic system as a whole, particularly for such elements of the system as the bend.

Description of advantages of the invention "PNEUMATIC INSTALLATION"

EFFECTIVINESS:

Pneumatic installation is used for removing of ferromagnetic chip and iron dust from the cutting zone of machine tool by pneumatic method mainly for the stainless and high-strength steels.

The other main objective of this device is the breaking of chip which is formed in process of the cutting and naturally makes a lot of problems for the machinist and machine tool because has a very long length and high temperature.

NOVELTY:

In pneumatic system is arranged the new element in view of plasmatron which is fixed between the dust-chip arrangement and cyclone and has two carbon rods supporting in process of transportation of the ferromagnetic material the high voltage arc.

And besides the plasmatron has also the movable element in view of plain matrix which has many holes so that frontal surface of this matrix is done in view of concave and its down part of said plasmatron on the way movement of this material is installed the water (sprinkler or spray) system for cooling of this material.

UTILITY AND ADVANTAGES:

In process of cutting the detail forms the chip which catches by the dust-chip arrangement and then comes into zone of arc of plasmatron where this chip melts and moves further through the holes of the matrix, forming the granules (spheres) by means of cooling its in water system.

By that in zone of plasmatron the chip receives form of granule (convention named as sphere) and naturally at this variant changes the coefficient of friction—primary element chip in dust-arrangement has the friction of slipping and later after of the matrix this material (chip as sphere) has the friction of rolling and this fact decreases in total the demanded power for the pneumatic transportation of material chip and improves the technological process in whole.

7. Magnetic-dynamical pipeline

Referring to the other invention which is shown in Figure 1 and 2, we see the main essential of this device called the Magnetic-dynamical pipeline.

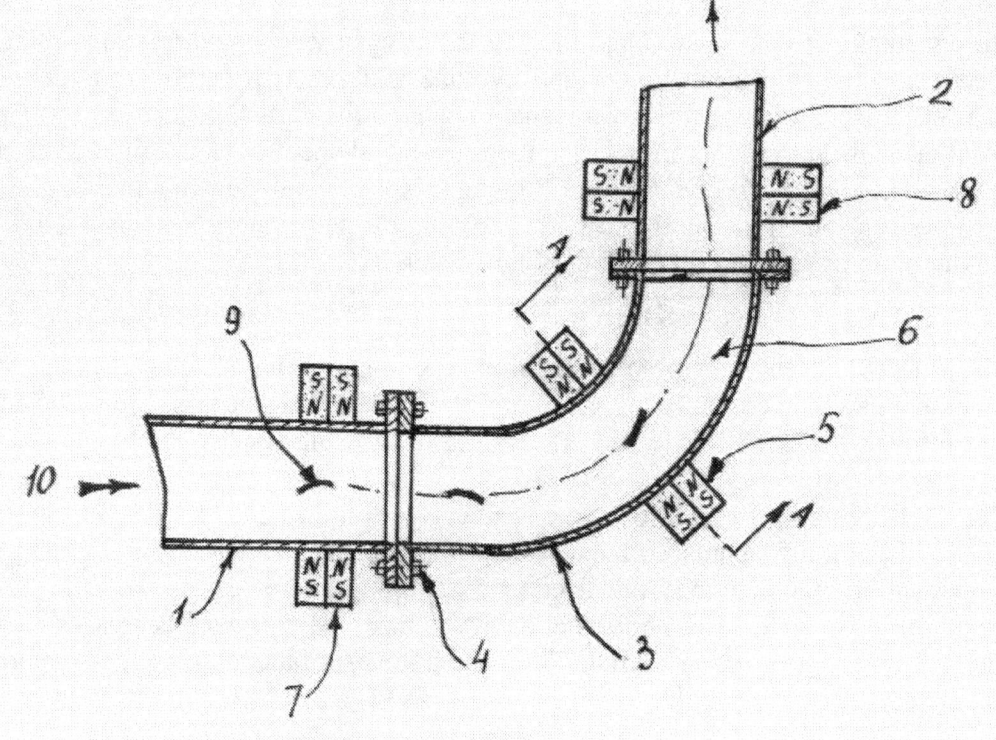

A.ROZENBLAT
MAGNETICODYNAMICAL PIPELINE

FIG.1

Anatoly Rozenblat
INVENTOR

A.ROZENBLAT
MAGNETICODYNAMICAL PIPELINE

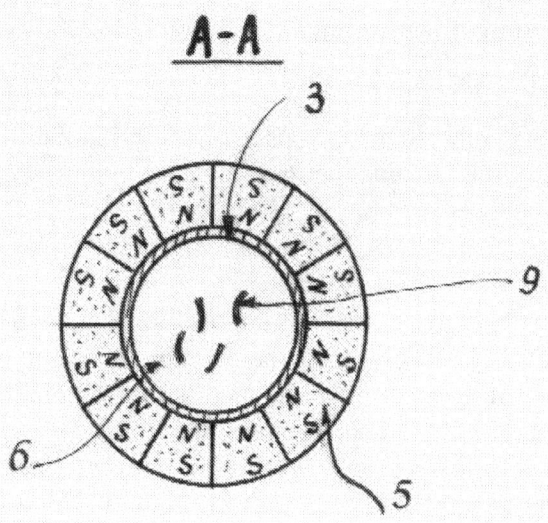

A-A

FIG.2

Anatoly Rozenblat
INVENTOR

NON-CONFIDENTIAL INVENTION SUMMARY

AIR # 300

MAGNETIC-DYNAMICAL PIPELINE

(Inventor: Anatoly Rozenblat)

INVENTION DESCRIPTION:

This device is used to decrease the abrasion wear of elbows and pipeline surfaces in period of transportation of the ferromagnetic materials.

UTILITY AND ADVANTAGES:

The pipelines and elbows widely use for the pneumatic transportation of the different materials particularly the ferromagnetic materials in view of chip and dust from the machine tools in cutting processes.

Such metal particles have a very high abrasion index, causing continual wear and damage to the pipeline which transport it.

This damage is most pronounced at the elbow and curves of the pipelines when the differential inertia of the particles ends them against the pipeline wall.

This invention utilizes a non-metallic element of the pipeline, combined with a magnetic field created by magnets of the same poles it are situated oppositely one to other.

Magnets creates a magnetic field which presents the repellent forces so that the ferromagnetic material moves non-contacting and not—interacting with the inner surface of elbow and pipeline reducing the total contact with surfaces of the resulting abrasion.

PATENT AND DEVELOPMENT STATUS:

This innovation has been filled and protected by the author in U.S Copyright in 1991.

FURTHER INFORMATION:

Copies of this innovation are available upon request. Data demonstrating its efficacy of use available from the inventor under cover of a Confidential Disclosure Agreement. Commercial rights are available for exclusive, non-exclusive or field-of-use exclusive licensing. When making inquiries, PLEASE REFER TO THE AIR # SEEN ABOVE.

8. Helicopter

The present innovation takes primary addresses the air technique of helicopters and other flying apparatus and is designed for promotion of speedy movement and improvement of the maneuvering characteristics of helicopters. In Figures 1,2 and 3 we see the essence of the present invention.

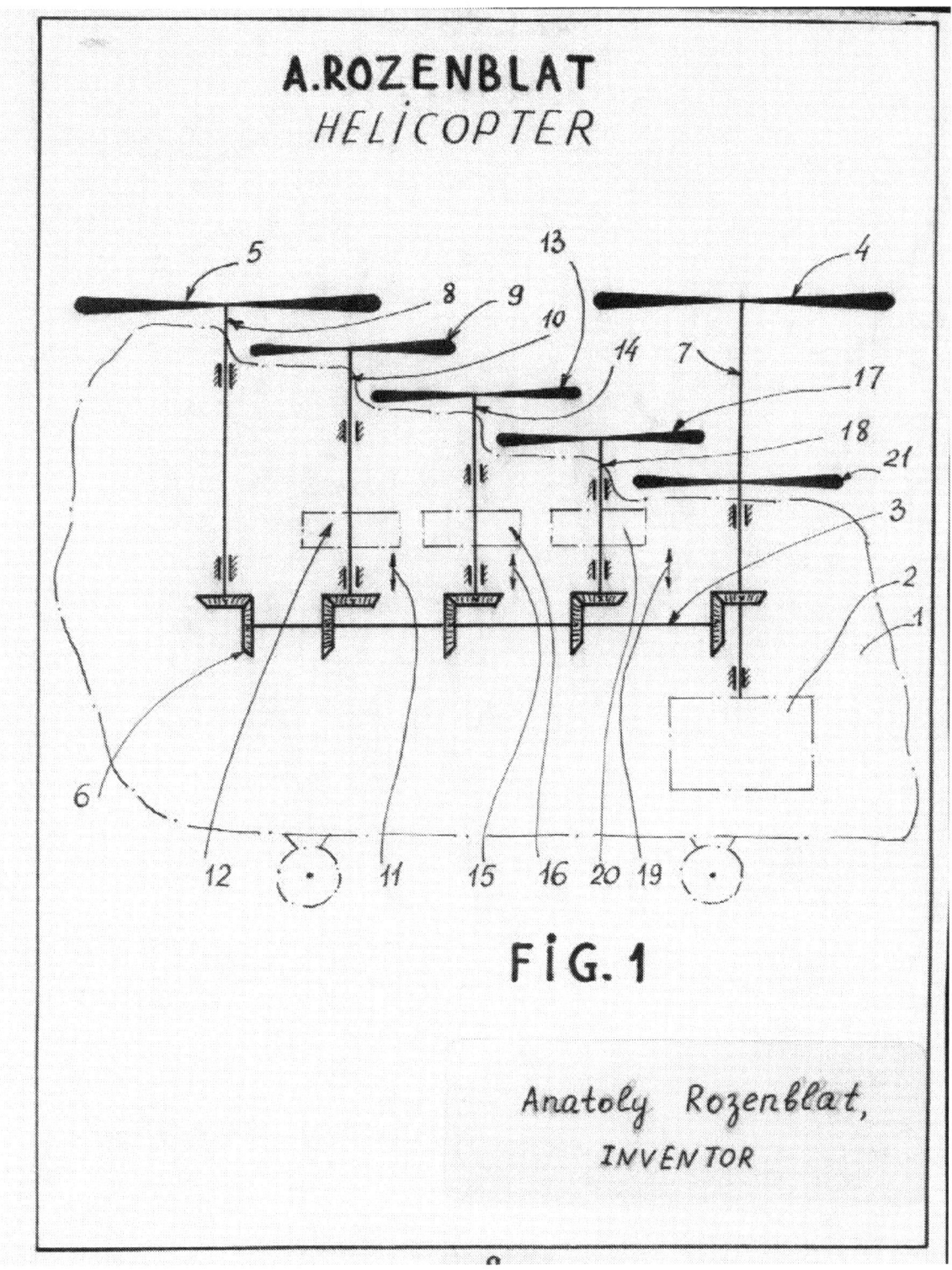

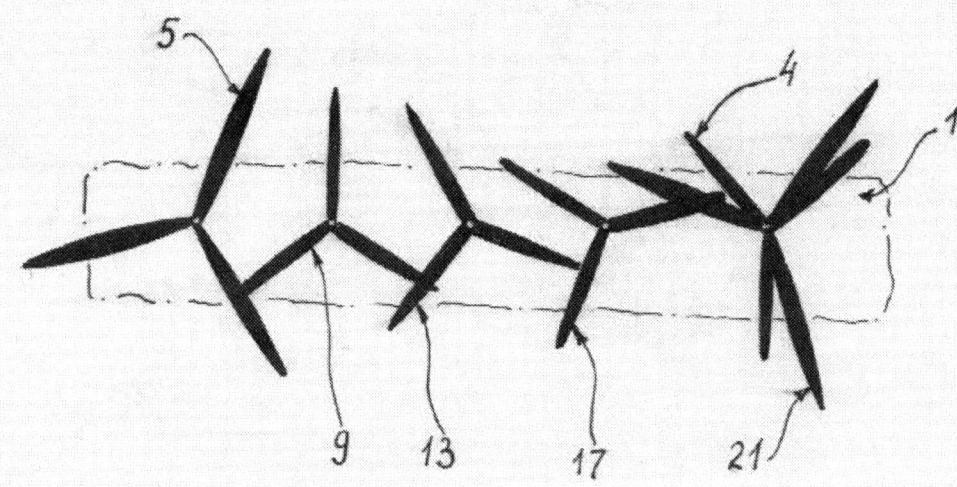

FIG. 2

84

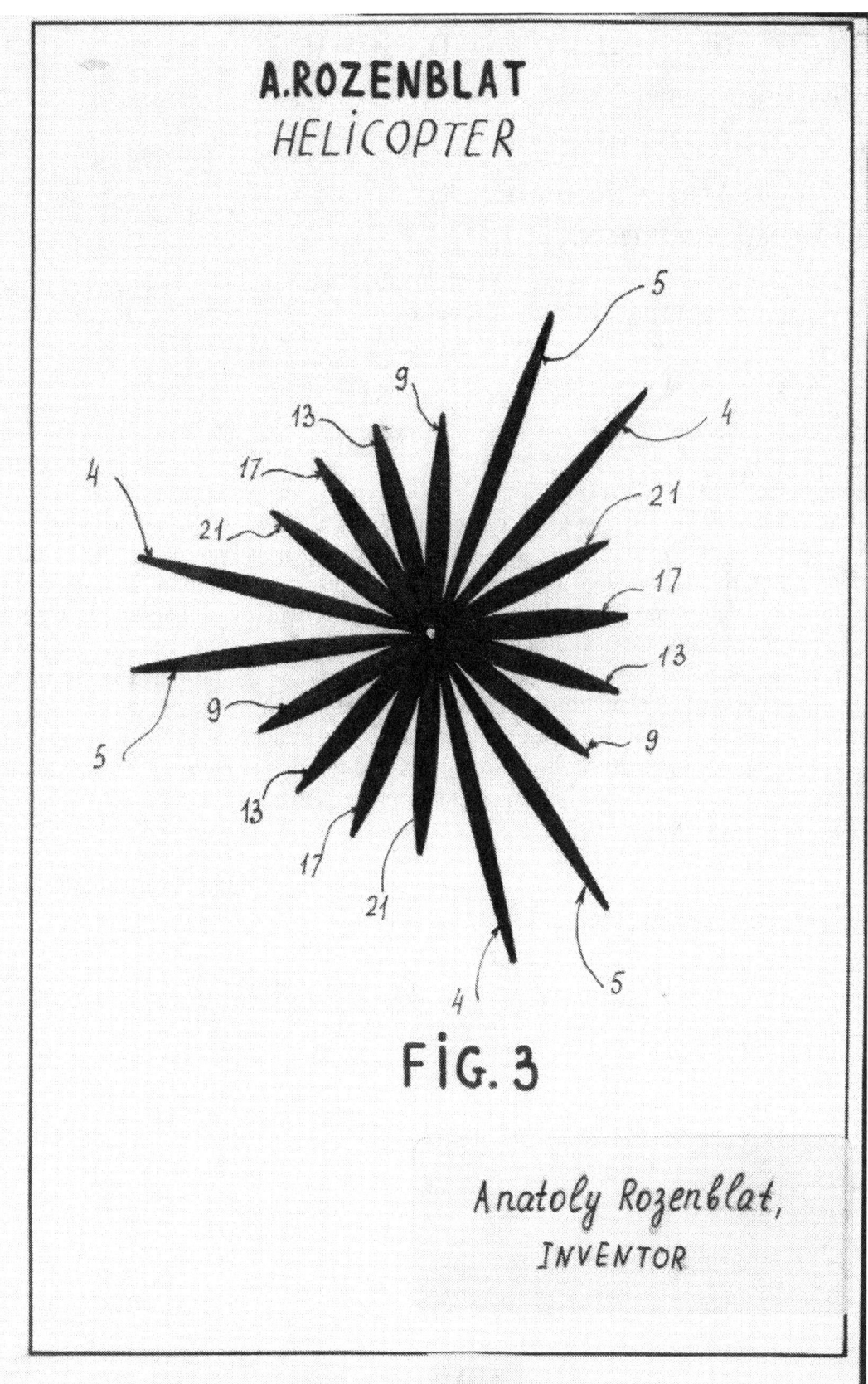

FIG. 3

NON-CONFIDENTIAL INVENTION SUMMARY

AIR #301

HELICOPTER

(Inventor: Anatoly Rozenblat)

INVENTION DESCRIPTION:

This device is used to guarantee security of flight on occasion breakage the blades of screws in period of flying and also to improve the maneuver-tactical characteristics of helicopter.

UTILITY AND ADVANTAGES:

The present helicopter does not guarantee the security of flight and has a small load-capacity.

This defect presents also on the helicopter with two basic screws it are situated oppositely to the end of fuselage given on certain distance one higher of other.

This invention utilizes the construction in view of n-quantity screws on the vertical and parallel axis of rotation having movement from the power installation. Diameter of each screw in whole determines by the minimum of meaning when commensurable with given distance and safety conditions of rotating screws.

But each additional screw has movable axis of rotation it has the back-progressive movement in the vertical plane from the auxiliary system, for instance hydraulic.

In the process of flying helicopter additional to the basic system of screws attach the system of bringing n-quantity screws which produce more summary air-dynamical flow it increases the speed of flight and improve the maneuver-tactical characteristics of helicopter.

PATENT AND DEVELOPMENT STATUS:

This invention has been filled out and protected by the author in US Copyright in 1991.

FURTHER INFORMATION:

Copies of the invention are available upon request. Data demonstrating its efficacy of use available from the inventor under cover of a Confidential Disclosure Agreement. Commercial rights are available for exclusive, non-exclusive or field-of-use exclusive licensing. When making inquiries, PLEASE REFER TO THE AIR # SEEN ABOVE.

SAFETY OF FLIGHT

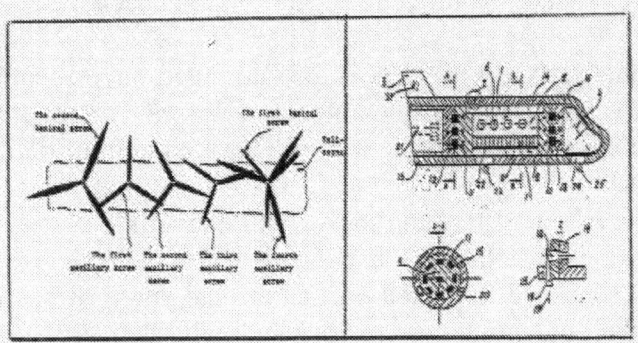

FIG. 16 SCHEME OF BLADES
DISTRIBUTION IN HELICOPTER

FIG. 17 AIRCRAFT WITH SEPARATION
OF PASSENGER SECTION

1. RELIABILITY AND DURABILITY OF SYSTEM
   DRIVE IMPROVES.

2. IN EXTREMAL SITUATIONS OF FLIGHT (FIRE,
   RANDOM FAILURE OF ENGINE) THE SAFETY
   AND LIFE OF PASSENGERS PROVIDES.

3. SPECIAL DESIGNED SYSTEM GUARANTEES THE
   SPEED DISCONNECTION OF PASSENGER
   SECTION.

The helicopter 1 includes the engine power 2 and the system of delivery 3 for rotating the first 4 and the second 5 basic screws. The system of delivery 3 is completed by the gear transmission 6which receives its rotation from engine power 2.

From Figure 1, we also see that the two basic screws 4 and 5 are installed accordingly on the vertical and parallel axis 7 and 8 with the distance given. On the helicopter 1 four screws are installed additionally.

The first screw 9 with the blades is installed with the displacement in the vertical plane according to the basic screw 5 that secures the safety of rotation and is fixed on the vertical shaft 10. The shaft 10 with the screw 9 has the back-progressive movement 11 from the autonomic system (or the hydraulic delivery system 12) which gets into smooth connection with the system of delivery 3 in case of breakage or steep blades of the screw 4 and 5 or the need to increase the speed of movement and load-capacity of the helicopter while maneuvering.

The other additionally fixed screw 13 is also arranged on vertical shaft 14 and has the back- progressive movement 15 from the autonomic system (the hydraulic delivery system 16), and also has the smooth connection with the basis system of delivery 3 in case of the above-named variants.

The blades of the screw 13 accordingly with the other screw 9 also displace and naturally this screw 13 is installed so that it guarantees the safety of rotation. The third additionally installed screw 17 is placed on the vertical shaft 18 has the back-progressive movement 19 from the autonomic system (the hydraulic delivery system 20) and has the smooth connection with the basis system of delivery 3 in case of the scenarios described above.

The fourth screw 21 is installed on the shaft 7 of the basis screw 4 and has the rotation from the engine power [the basic screw 5 has rotation from the system of delivery 3], and also the screw 21 is fixed so that it has displacement relative to the third screw 17.

The outer diameter of the screw 21 guarantees the safety of rotation. Figure 1 shows that the screw 21 is fixed rigidly on the axis of rotation of the basic screw 4, but in keeping with the previous screw 17, the screw 21 is displaced and installed with the screw 4 on the same vertical shaft 7.

As we see in Figure 2, the helicopter 1 has six screws: two bases screw 4 and 5 and four other screws 9,13,17 and 21. From Figure 2, we see that the blades of each screw relative to the next are displaced by a value equal to the formula:

$\phi = I \times (360/ KN)$ (1)
where,
360= the value of full angle of the rotating blade of the screw, in degrees;
I = the current number of screw;
K =quantity of blades that are installed on the screw;
N = total quantity of screws which are arranged on the helicopter.

In Figure 3, the distribution and installation of the blades of the screws in an angle system of coordinates is shown schematically.

As a sample, the author offers some calculations, for installation and distribution of each screw accordingly to Figure 3.

*Basic screw 4 (the blades 4-4-4) is situated:
$\phi = I \times (360 / KN)$; $I=0$; $K=3$ $N=6$ then $\phi = 0$

- The screw 21 (the blades 21-21-21) is displaced:
  $\phi = 1 \times (360/18) = 20°$
  *The screw 17 (the blades 17-17-17) is displaced:
  $\phi = 2 \times (360/18) = 40°$
- The screw 13 (the blades 13-13-13) is displaced:
  $\phi = 3 \times (360/18) = 60°$
- The screw 9 (the blades 9-9-9) is displaced:
  $\phi = 4 \times (360/18) = 80°$
- The screw 5 (the blades 5-5-5) is displaced:
  $\phi = 5 \times (360/18) = 100°$

In the process of using the helicopter the two basic screws the helicopter takes as the base are the screws 4 and 5. But in an emergency situation or other variants in the period of movement from one place to another when the helicopter needs a higher speed of movement or it becomes necessary to increase the load capacity and to improve the maneuvering characteristics of the helicopter it is possible to install the n-quantity of screws on this invention.

In this case, the new screws 9,13 and 17 would be included in operation and would form a more efficient air-dynamic flow for the helicopter in whole. In case of breakage of the basic screws 4 and 5 other screws 9,13 and 17 can be used that guarantee the safety of flying.

From this conclusion, we can see that the present invention guarantees a higher load capacity for a helicopter and increases the velocity of the helicopter's movement because the helicopter has additionally arranged and fixed the n-quantity of screws with the blades.

This also gives a more air-dynamical flow and guarantees the safety of flying in case of breakage of the basic screws and also decreases the drain of fuel resources in result of the reduction of the quantity of helicopters in whole for the industry.

9. Seismic combined nuclear system

The seismic combined nuclear system relates to high technology and guarantees the safety of nuclear power stations and the environment in the event of an earthquake. The present invention is used for increasing the security of a nuclear system and the life of people. Referring now to the Figure 1 and 2, we see the general essentials of the invention.

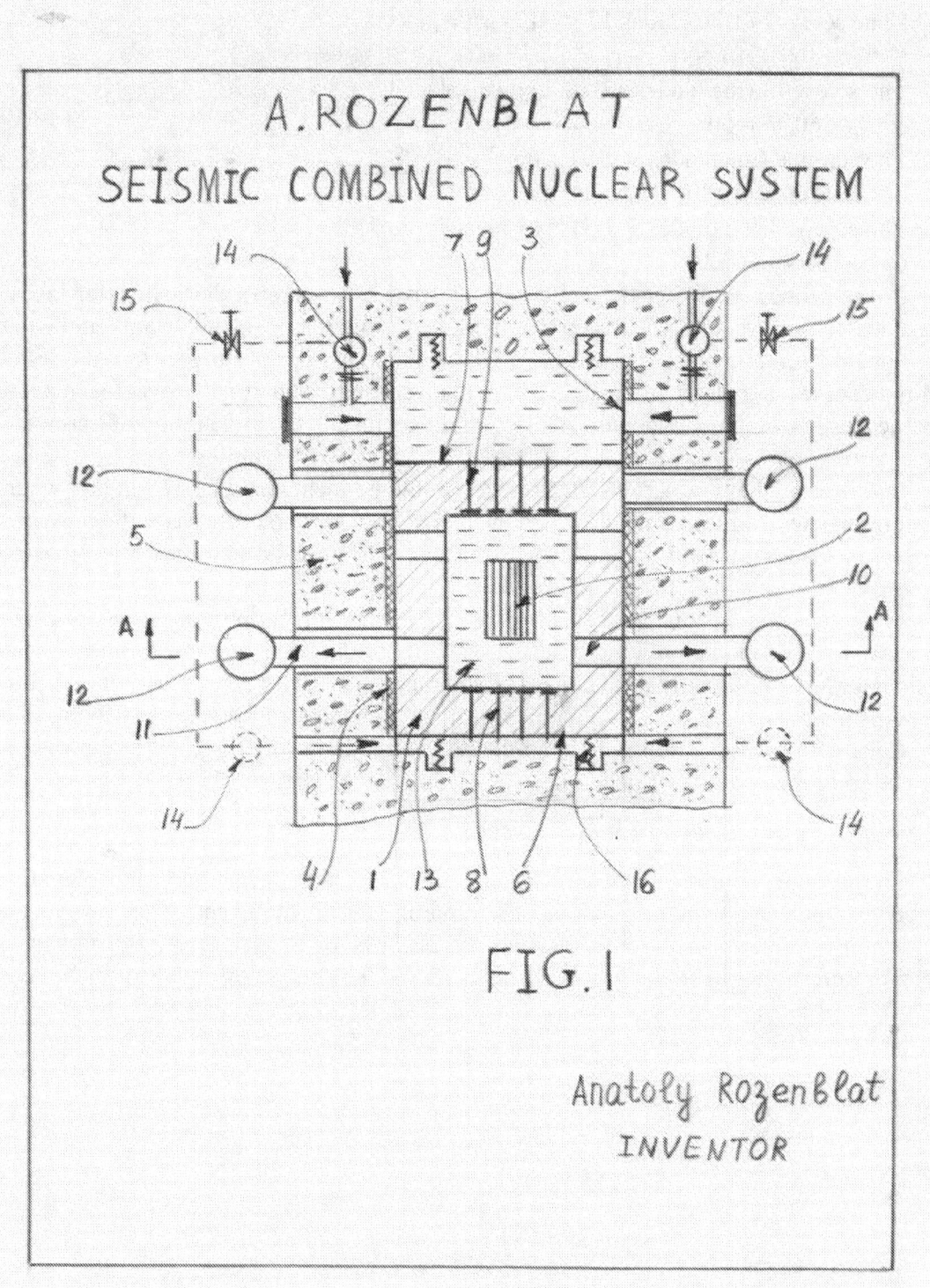

A.ROZENBLAT

SEISMIC COMBINED NUCLEAR SYSTEM

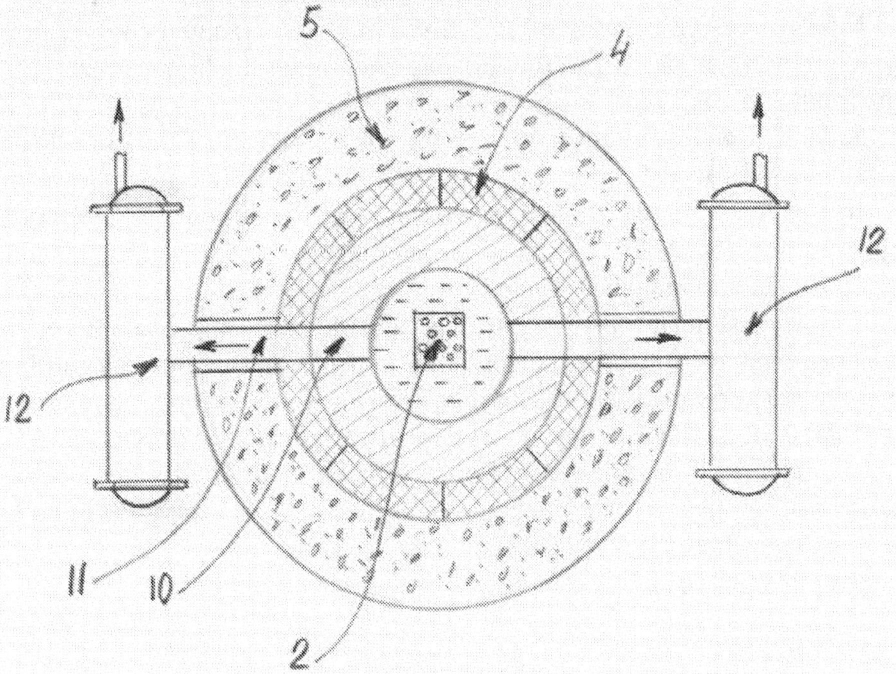

FIG.2

Anatoly Rozenblat

INVENTOR

Nuclear power station consists of a reactor vessel 1 with a core 2 which is in view of a piston arranged horizontally, and has on the perimeter of the cylinder 3 the bimetal surfaces 4. This surface 4 contacts with the piston 1(reactor vessel) and has the outer surface 5 in view of concrete material for the safety of system in total.

On both the lateral sides 6 and 7 of the piston 1 the special installation valves 8 and 9 are situated. The piston 1 also has the multiple exhaust channels 10 which are joined with cavities 11 of the cylinder 3 and other elements—for example with the steam generator 12— for movement of the coolant system 13 from the main coolant pump 14 which is joined with the rest of the system by means of electromagnetic valves 15.

For security of movement of the piston 1 there are some buffer device 16. In the working process of the primary nuclear system, the coolant water enters the system by means of a main pump 14 in a channel 17 with the following characteristics: a coolant pressure of 2,250 psi (15.5 Mpa); a coolant velocity of 15.5 ft/sec (4.7 m/sec); and a coolant flow rate (13x106 lb/hr) which moves the piston 1 to the left side and, under the high pressure of coolant water, opens the input valve 9.

So, the coolant water further comes to the inner channel of the core 2 (at this point the valves 8 are closed). In the process of movement of the piston 1 as shown in Figure 2, the exhaust channel 10 is joined with the cavities 11 of the cylinder 3, and coolant water at a high temperature (617°F/325°C) comes through to the steam generator 12 and moves on the next steps provided in the nuclear system.

At the movement of the piston 1 to the right position, the coolant water comes to the electromagnetic valve 15 or uses the coolant pump 14, and the process continues. It must be noted that the reactor vessel 1 does not join firmly with the steam generators 12 and for this reason the safety of the nuclear system in case of earthquakes is guaranteed.

Also, the reactor of the nuclear system is situated horizontally and does not join firmly with the system, particularly in a seismic conditions when ground may be broken and for this reason this new method provides and advantages and utility.

The advantages of the nuclear plant are shown separately below.

INVENTIONS FOR HIGH TECHNOLOGY

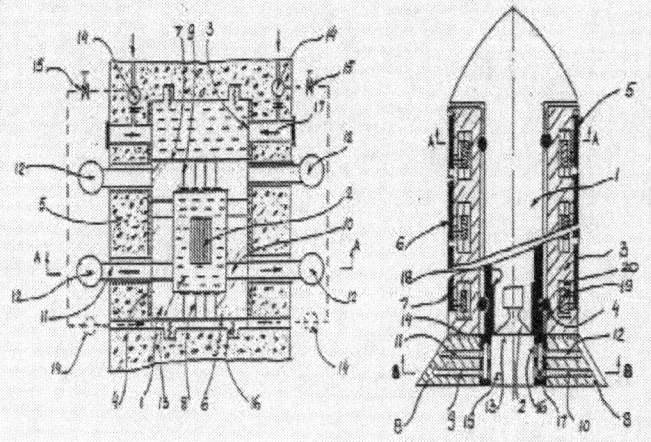

FIG. 19 ILLUSTRAION OF NUCLEAR PLANT (PWR)

FIG. 20 SCHEME OF COMBINED ROCKET "AIR"

ADVANTAGES:

1. EFFECIENCY INCREASES.

2. SAFFETY IMPROVES.

3. LIFE OF PEOPLE AND ENVIREMENT IMPROVES.

DESCRIPTION OF ADVANTAGES OF THE INVENTION:

"SEISMIC COMBINED NUCLEAR SYSTEM"

EFFECTIVENESS:

This device is used for nuclear power plant and industry. The present well-known nuclear systems do not guarantee the safety of nuclear power station as they are situated on the solid ground firmly and exposed to the seismic destruction.

And besides pressured—water power reactors (PWR) is situated also firmly and vertically that demands a high building for this installation.

These defects are absented at present device which used for increasing of the security nuclear system in whole.

NOVELTY AND ORIGINALITY:

This nuclear system consists from the reactor vessel with core, steam generator, main coolant system and other elements which are joined in one energetic system.

The new elements of this device there are that whole system is fulfilled horizontally and made in view of cylinder and piston, so that function of said movable piston makes up the reactor vessel with core embracing on this circle by combined cylinder having inner bimetal surface which is contacted with said piston which has the multiple exhausted channels on the peripheral surfaces for joining with the cavities of said cylinder and other elements for movement the water coolant in working processes.

UTILITY AND ADVANTAGES:

The reactor of nuclear system is situated horizontally and is not joined firmly with the system particularly in seismic condition when ground may be broken and naturally to form the splits.

PROFITABLE:

This innovation provides the security nuclear system and life of people and also guarantees the safety of environment from damage in case of earthquake.

10. Aircraft-passenger and method of its flight

The problem of ensuring safety for the people in civil and military aviation is a major problem for many countries. The improvement of conditions of flying for passengers is suggested by author in a new passenger aircraft which is shown in Figures 1, 2, 3, 4, 5, 6, 7, 8, 9 and 10.

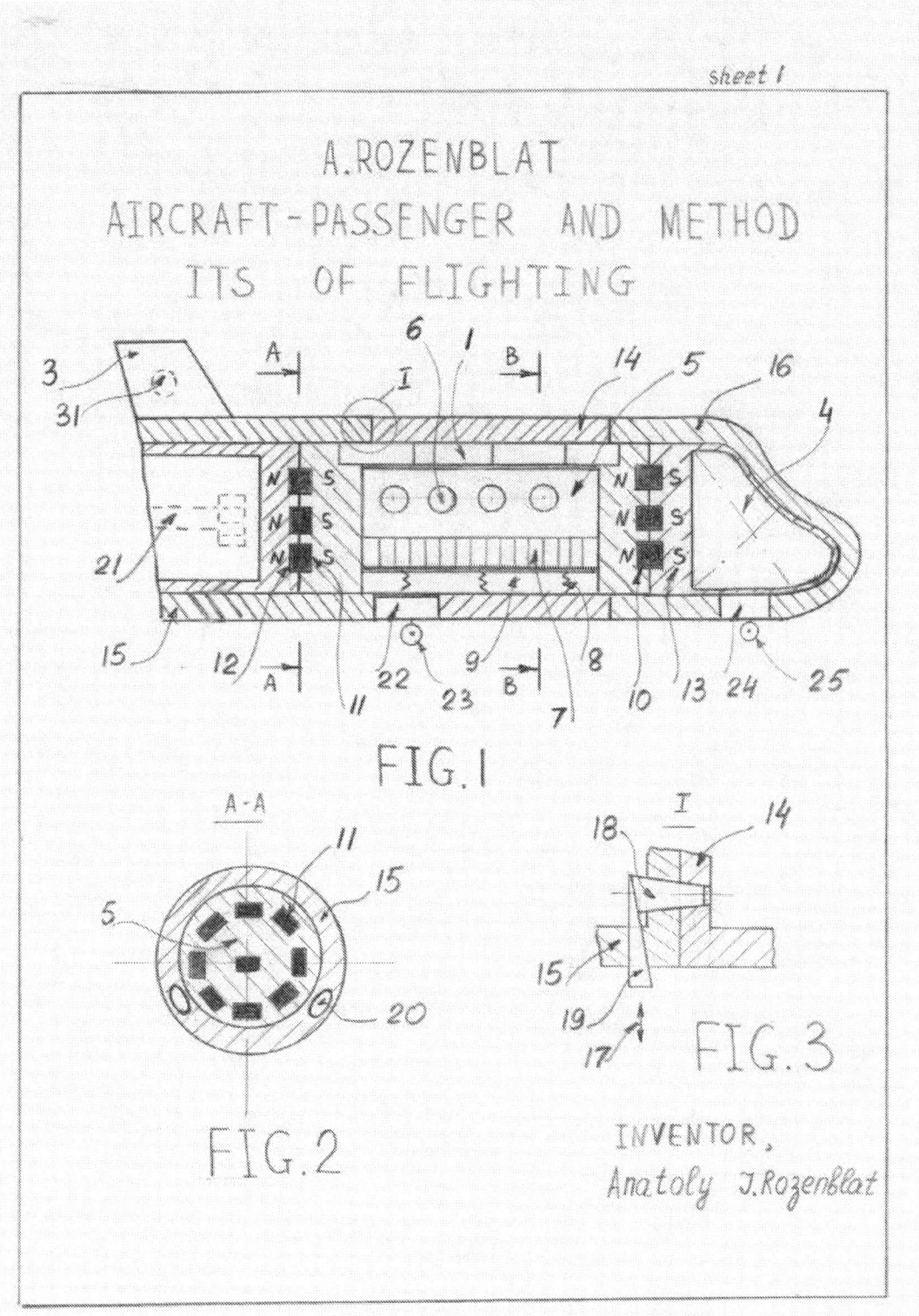

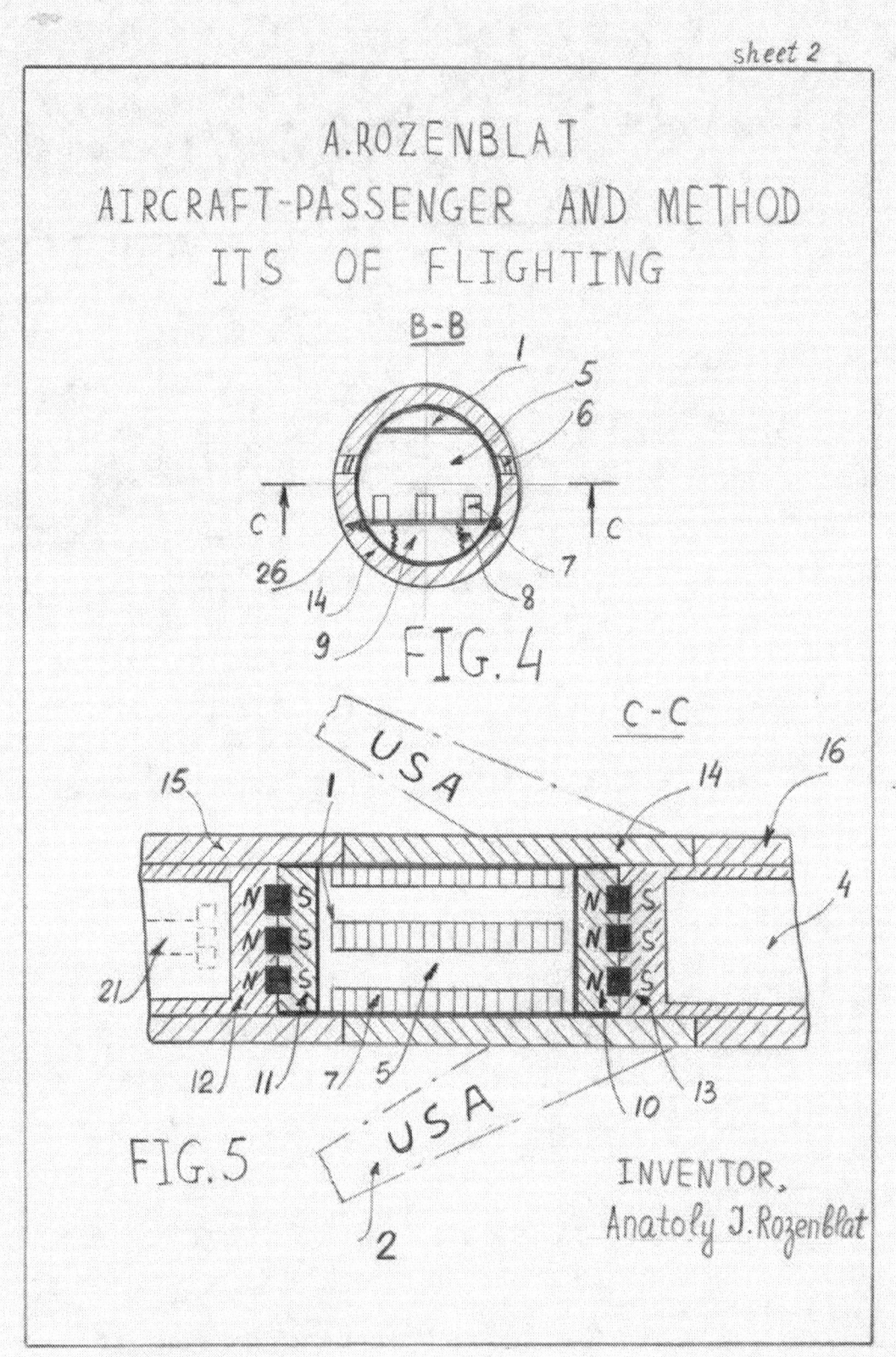

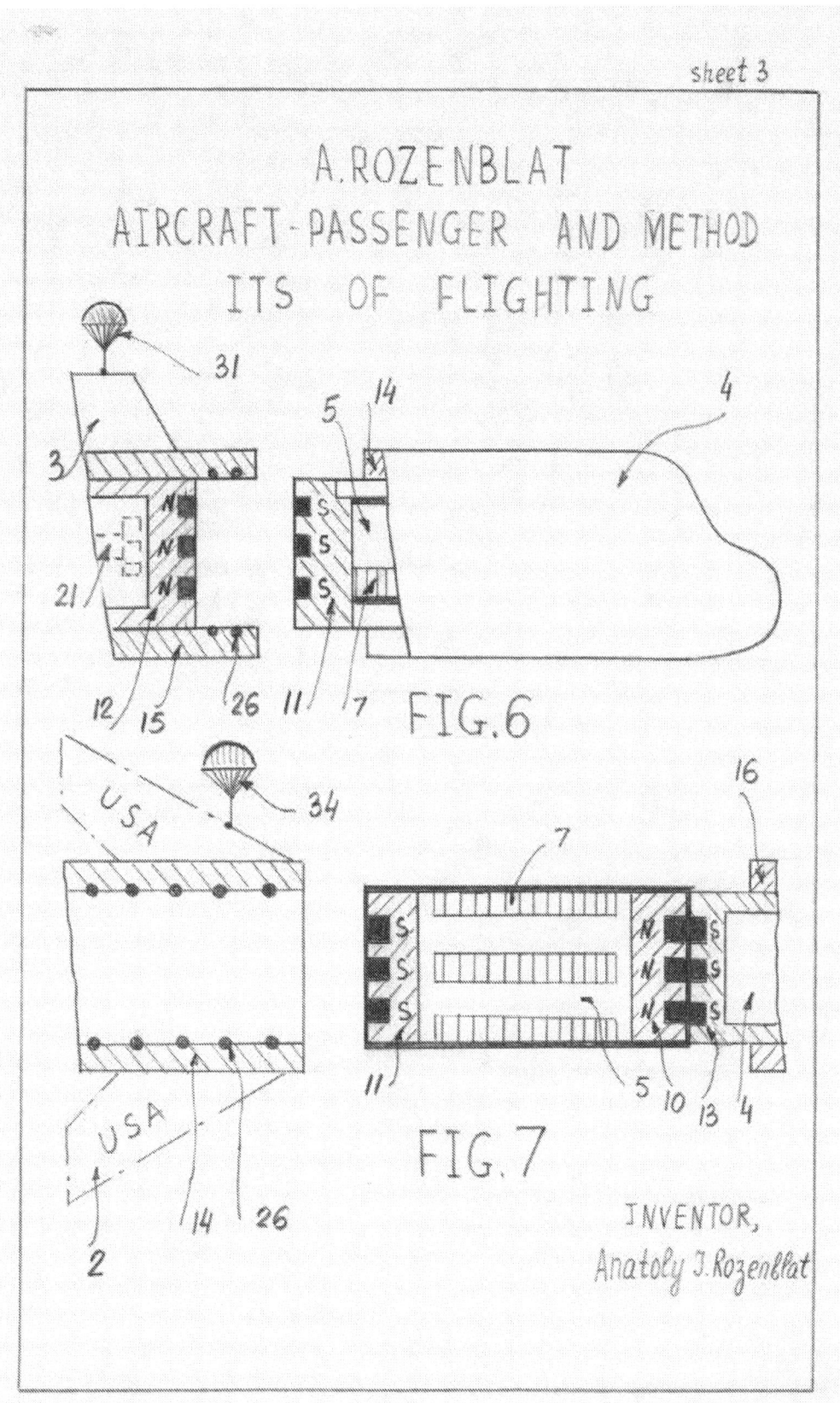

FIG.6

FIG.7

INVENTOR,
Anatoly J.Rozenblat

97

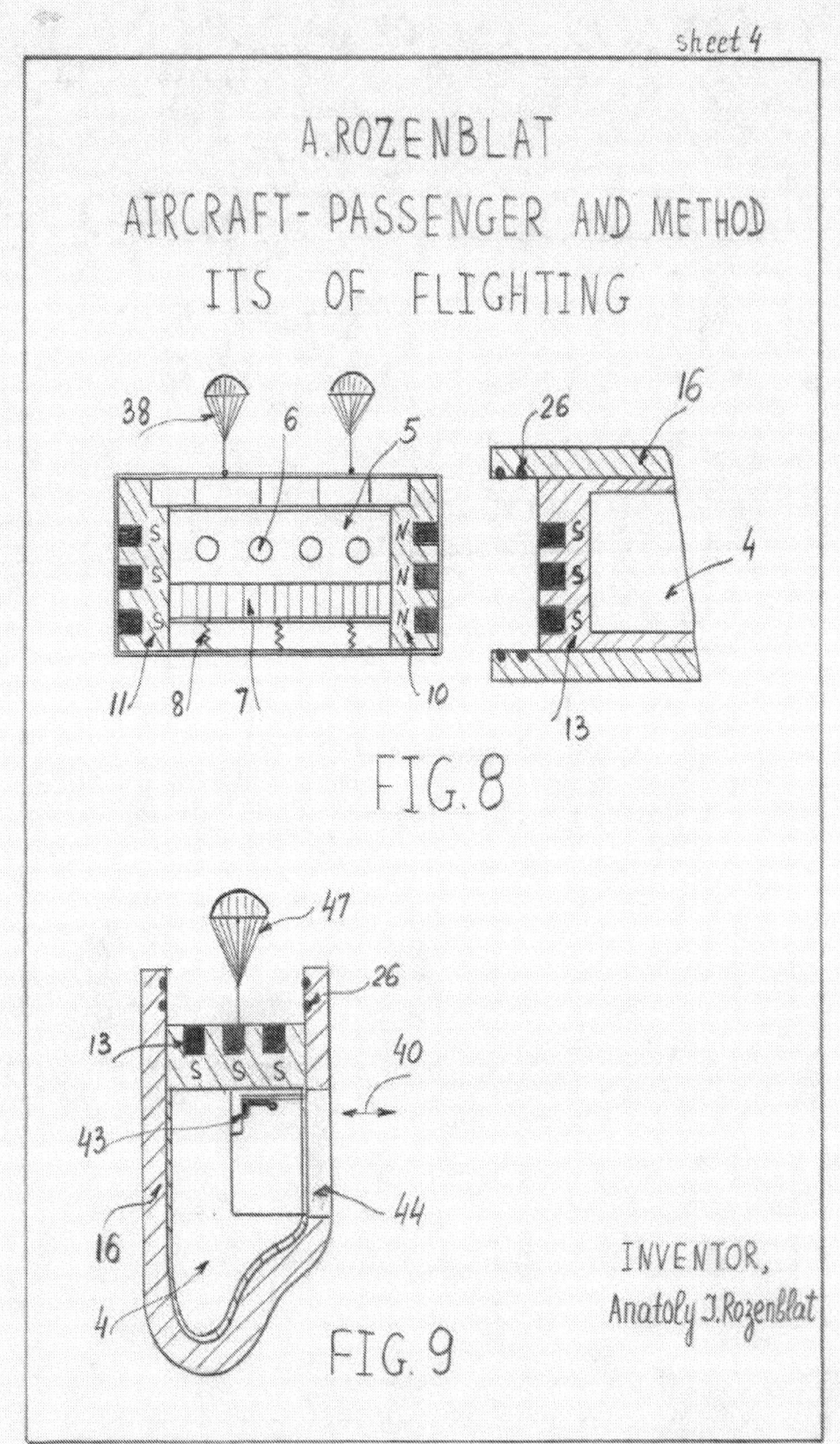

FIG. 8

FIG. 9

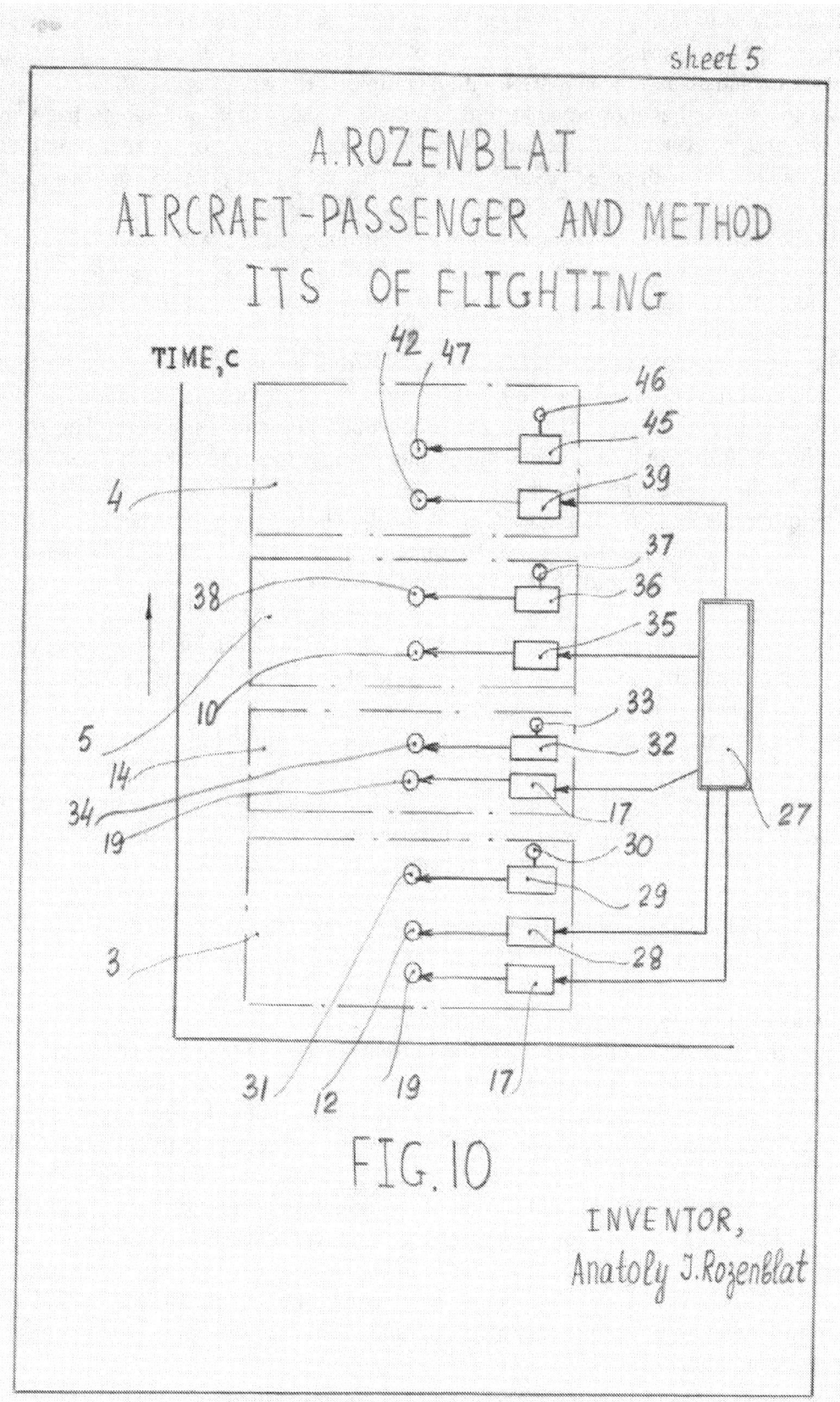

FIG. 10

Referring now to Figure 1, we see the general essential nature of this innovation. The passenger aircraft consists of the elements of the fuselage 1 with wings 2 (not shown) the tail system 3 and also the front part 4 with a complex driving system.

The fuselage 1 has the movable inner element 5, passenger hall, which has windows 6, seats 7, and a system of sections 8 and other holes 9 for wiring communication. Electromagnets 10 with pole "N" and electromagnets 11 with pole "S" are installed on both sides of movable inner element 5, and are joined with the tail part 3.

On the tail system 3 additionally more electromagnets 12 with pole "N" are installed, and on the front part 4 more electromagnets 13 with pole "S" are installed. The movable inner element 5 of the fuselage 1 embraces the other element 14 of the fuselage that has two wings 2 firmly fixed.

The outer element 14 is completed separately from the tail system 3 and the front part 4, on which are arranged the outer elements 15 and 16. These outer elements 14,15 and 16 are joined to the hydraulic system 17 by means of other elements 18 and 19. The first element 15 has holes 20 for connection with the engine 21, and the outer movable element 14 has a system of a chassis 22 with wheels 23.

The other element 16 has the system 24 for the chassis with the wheels 25. On the inner surfaces of the elements 14,15 and 16 the bearings 26 are installed. While flying in a dangerous situation, the pilot must switch on the automatic program of the system 27, as shown in Figure 10.

The tail part 3 with the engine 21 separates, and in this situation the hydraulic system 17 moves the element 19, and accordingly a second element 18 also moves in this case (i.e the entire system is open). Then the system 28 does not give the current to the electromagnets 11 and 12, and for this reason the tail part 3 separates from the aircraft, as shown in Figure 6.

In the process of dropping the tail part 3, the switch 29 receives the signal from the installation of time 30, which is joined with the parachute system 31 for the tail part 3. The system 17 (this is the separation system) moves the elements 18 and 19 the outer movable elements 14 and 16.

At this point, the fuselage (outer element 14) and the wings 2 moves from fuselage (inner element 5) and separates from the aircraft. In the process of falling, the fuselage (outer element 14) gives the signal to the switch of the parachute system 32 from the installation of time 33 directly on the parachute 34.

Then the system 35 does not give the current to the electromagnets 10 and 13, and for this reason the fuselage (movable inner element 5) as shown in Figure 8, which separates from the aircraft. In the process of falling, the fuselage (inner element 5) gives the signal to the parachute system 36 from the installation of time 37 directly on the parachute 38.

Finally, the front part 4 separates from the aircraft, and at this point, the system 39 receives the signal on the catapult switch of the direction 40 and the system 41 (not shown) of the parachute 42 for the pilot 43 through the window 44, as shown in Figure 9.

In the process of falling, the front part 4 of the aircraft gives the signal on the system 45 from the installation of time 46 directly on the parachute 47. So, we see that all systems of the aircraft separate in the process of flying and guarantee the safety for the passengers in emergency situations.

The advantages of this aircraft-passenger are shown below.

THE 21ˢᵀ CENTURY AIRCRAFT

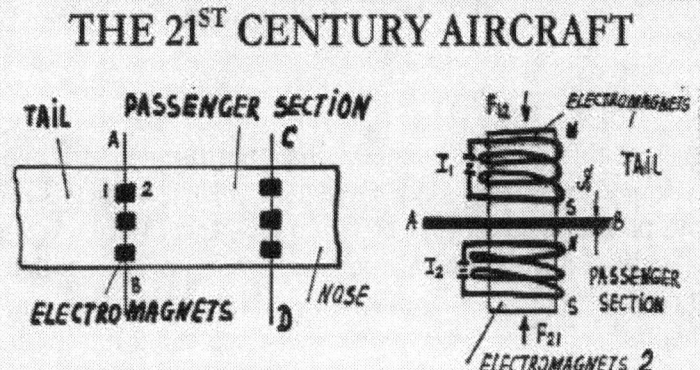

FIG. 15 SCHEME OF DESIGNING

Example:

1. THE ELECTROMAGNETIC FORCE

$$F_{12}=F_{21}=(B^2 \cdot S_a)/2 \cdot \mu_o;$$

B=MAGNITUDE OF THE MAGNETIC FLUX DENSITY $B=I/S_a$

S_a=CROSS-SECTIONAL ARE OF PLATE AB, $S_a=0.0025^2 m$;

μ_o=PERMEABILITY OF THE MAGNETIC CORE, $\mu_o=1.2 \cdot 10^{-6}$;

2. THE MAGNETOMOTIVE FORCE mmf=N*I A-turns;
   $N_1 \cdot I_1 = N_2 \cdot I_2 = 3000$

3. RELUCTANCE OF EACH COIL, $R_1 = R_2 = 1.0186 \cdot 10^8$
   A-turns/wb;

4. LOOP CURRENT IN EACH COIL, $I_1=I_2=N \cdot I/R$; $I_1=I_2=3 \cdot 10^{-5}$ A
   SO, $F_{12}=F_{21}=62.5$ NEWTONS,   (B=5.0 TESLA)

5. TOTAL FORCE $\sum F_{12} = \sum F_{21} = F_{12} \cdot n$ (n=number of coil, n=100)
   $\sum F_{12} = \sum F_{21} = 6250$ NEWTONS=6.25KN

6. THE MECHANICAL FORCE ACTING BETWEEN TOTAL
   PAIR ELCTROMAGNETS 1 AND 2 IS:
   $\sum F_M = 12.5$KN

DESCRIPTION OF ADVANTAGES OF THE INVENTION:

"AIRCRAFT-PASSENGER AND METHOD ITS OF FLIGHTING"

USEFULNESS:

This device is used in civil and military aircraft. The present well-known aircraft does not guarantee the life of passengers in accident situation and also has many problems in question of transportation, for instance by sea or railroad.

These defects are removed at this device which improves the safety of flight for the passengers and crew and also guarantees the transportation it.

NOVELTY AND ORIGINALITY:

The new elements at this aircraft there are that the fuselage is made in view movable inner of element having on both lateral sides installed electromagnets so that they are joined with the front and tail parts of said aircraft which in total view form the opposite terminals.

And besides the method of flight the aircraft-passenger consists in that primary separates the tail part and then movable outer element of said fuselage with the fixed wings and movable inner element of said fuselage with the passengers and finally the front part said of aircraft with the crew.

PROFITABLE:

In this device provides by means of rescue of life the passengers and crew from aircraft in during of accident situation (broken engine, fire, etc.).

And besides separated elements of the aircraft could be used later for the connection of this aircraft on the next flight because the separation device uses the parachute system.

11. Combined rocket "AIR "*

The present invention is relevant to space and rocketry techniques (mainly to liquid-propellant rockets) and was designed to reduce the metal needed in a system, thereby improving the fuel economy and increasing the coefficient of useful action of the rocket as a whole. The combined rocket "AIR" is shown in Figure 1,2,3,4,5 and 6.

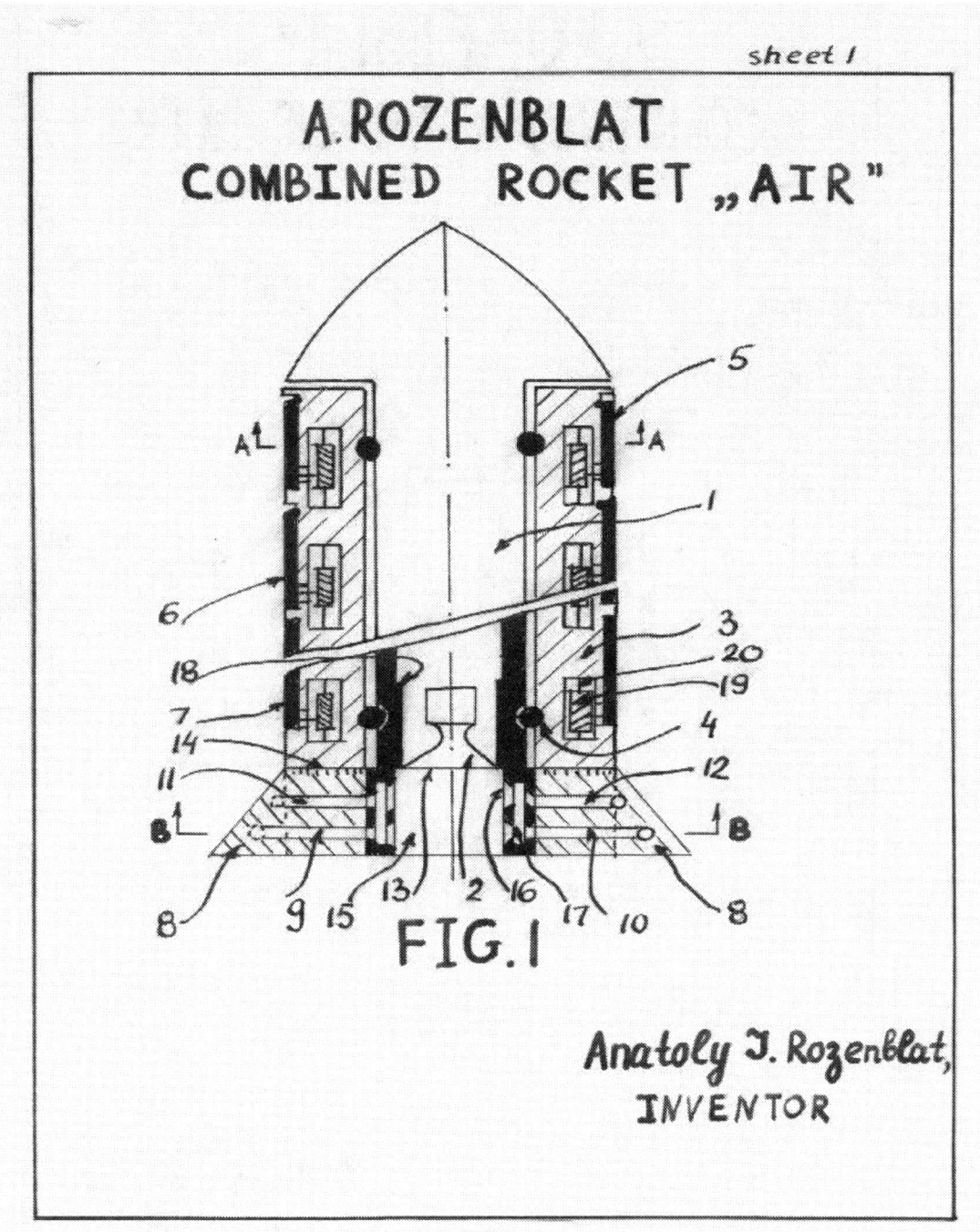

*Anatoly Isaacovich Rozenblat

A.ROZENBLAT
COMBINED ROCKET „AIR"

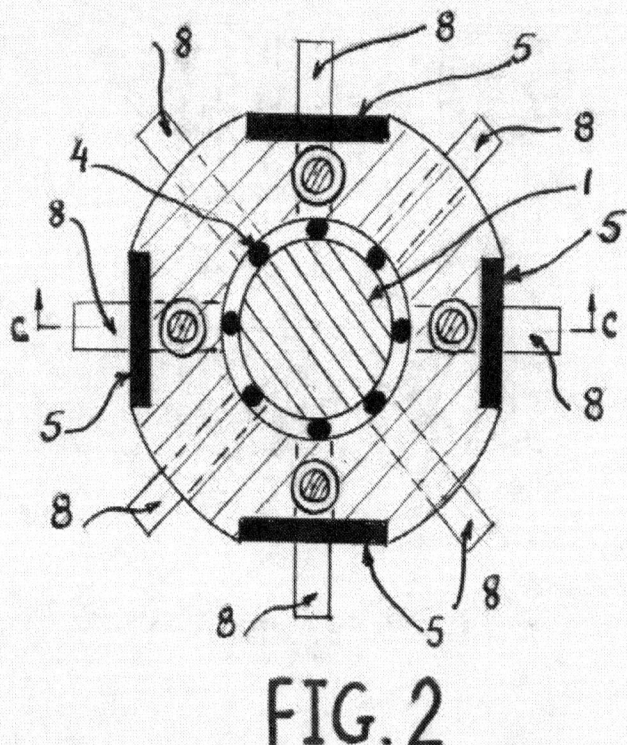

FIG. 2

Anatoly I. Rozenblat,
INVENTOR

A. ROZENBLAT
COMBINED ROCKET „AIR"

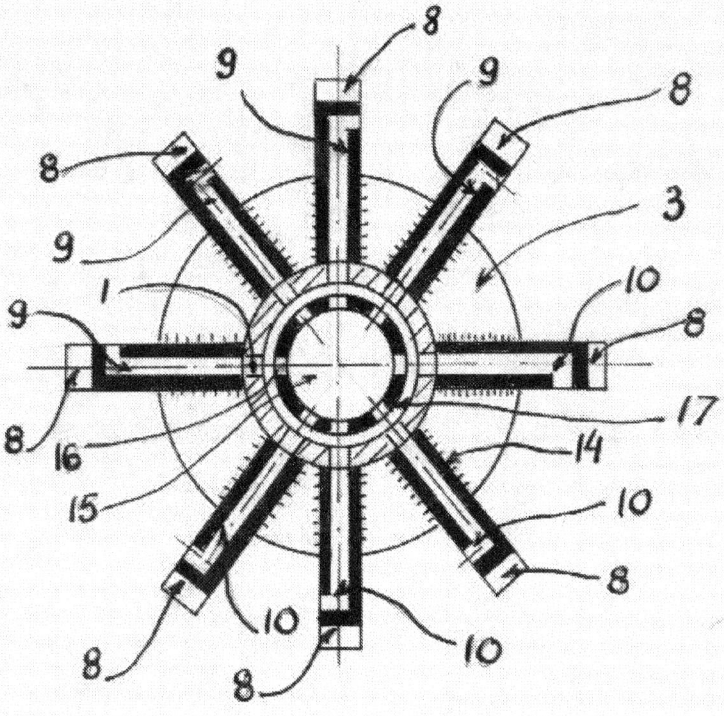

FIG. 3

Anatoly J. Rozenblat,
INVENTOR

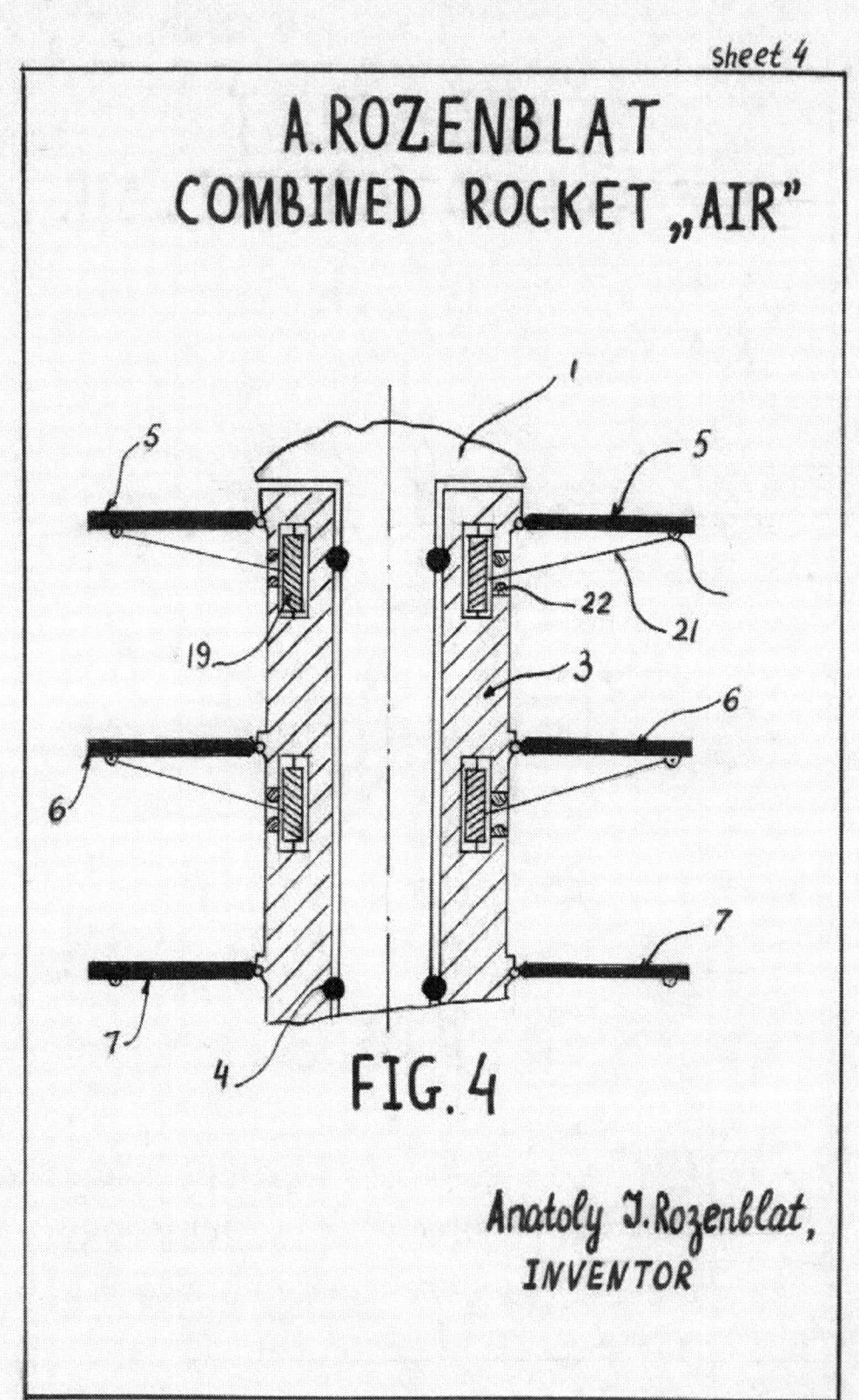

FIG. 4

A.ROZENBLAT
COMBINED ROCKET „AIR"

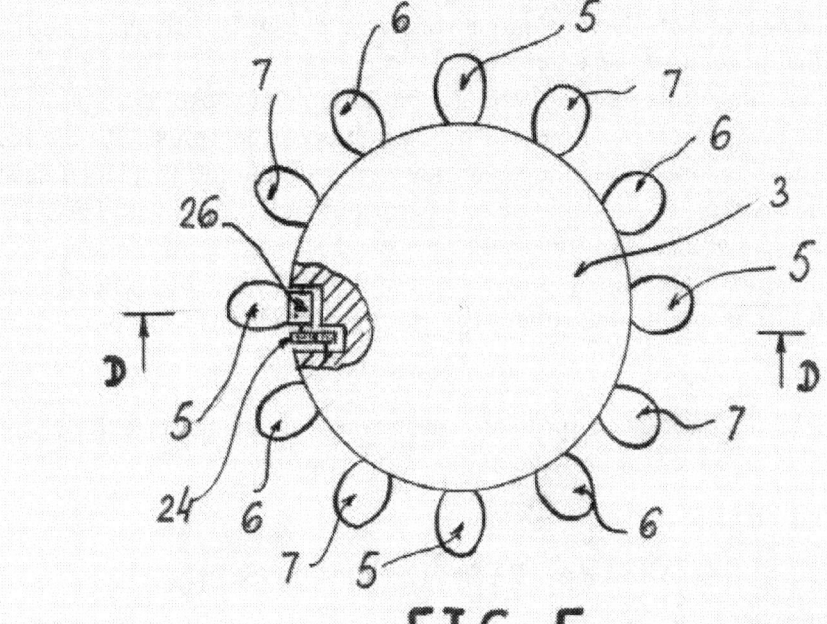

FIG.5

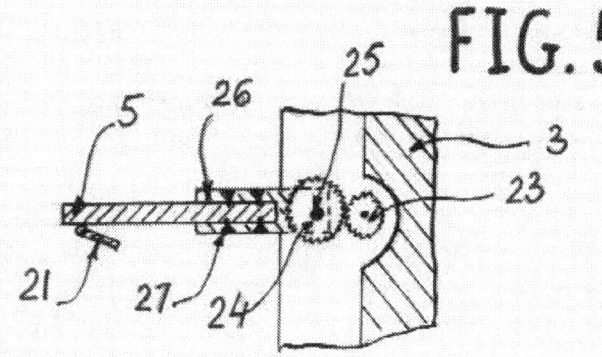

FIG.6

Anatoly J. Rozenblat,
INVENTOR

DESCRIPTION OF ADVANTAGES OF THE INVENTION:

"COMBINED ROCKET AIR"

USEFULNESS:

The present device has attitude to the space and rocketry industry mainly to the liquid-propellant rockets and used for the decreasing of metal-capacity system and fuel-economical components of rocket in whole.

NOVELTY AND ORIGINALITY:

The well-known rockets have a large weight and low coefficient useful action. This device is used that to remove these defects.

The new elements at this device there are that combined rocket body is made in view of inner motionless part which is connected through support bearings with the movable outer element which has the air-screw blades so they firmly fixed and displaced accordingly one to another on the level of the movable elements forming additionally power system.

And besides the stabilizers of this rocket are situated beneath of the level exhaust system engine and joined with additional power system so that movable outer element uses the exhaust gases from the engine and rotates the air-screw blades of this combined rocket.

PROFITABLE:

The profitable at this device provides by means that guarantees the economy of fuel components and also decreases the weight of rocket in whole.

12. Combined engine

ABSTRACT OF THE DISCLOSURE

A combined engine presents the system of two mechanisms: from one side this device consists from a crankshaft phasing engine for converting energy of thermodynamic process into mechanical energy such as transferring motion to mechanism of transmission having one or more movable axially piston whereby connected rod of said engine and also a motionless element cylinder embraced to said piston, and to another side this device has additionally new elements forming new energy on said engine-electric energy such as inducing an alternating current in a generator device at least having the permanent magnets rigidly mounted of said movable piston in the progressive direction and fixed on periphery arranging only surface with to said cylinder forming a magnetic field surrounding surface of said piston and installed the permanent magnets on lower part of said piston on side of crankcase space on said of cylinder and also elements of a coils inducing one or more separate coils connected a rigid unitary one with other in series circuit and installed inside of said cylinder on perimeter along his cylindrical surface and generatrix element of said cylinder to special made hollows in a form of said coil and their contacting surfaces with to said permanent magnets arranged with minimal space relative to said cylinder.

In process of committing a full working cycle in a crankshaft engine, a piston with the permanent magnets (they fulfill a function of solenoid) produce a magnetic field, which on Faraday's law, in a winding of excitation is installed in a hollow to said of cylinder (they fulfill a function as a inductance—coil), induces a electromotive force (e.m.f) and current.

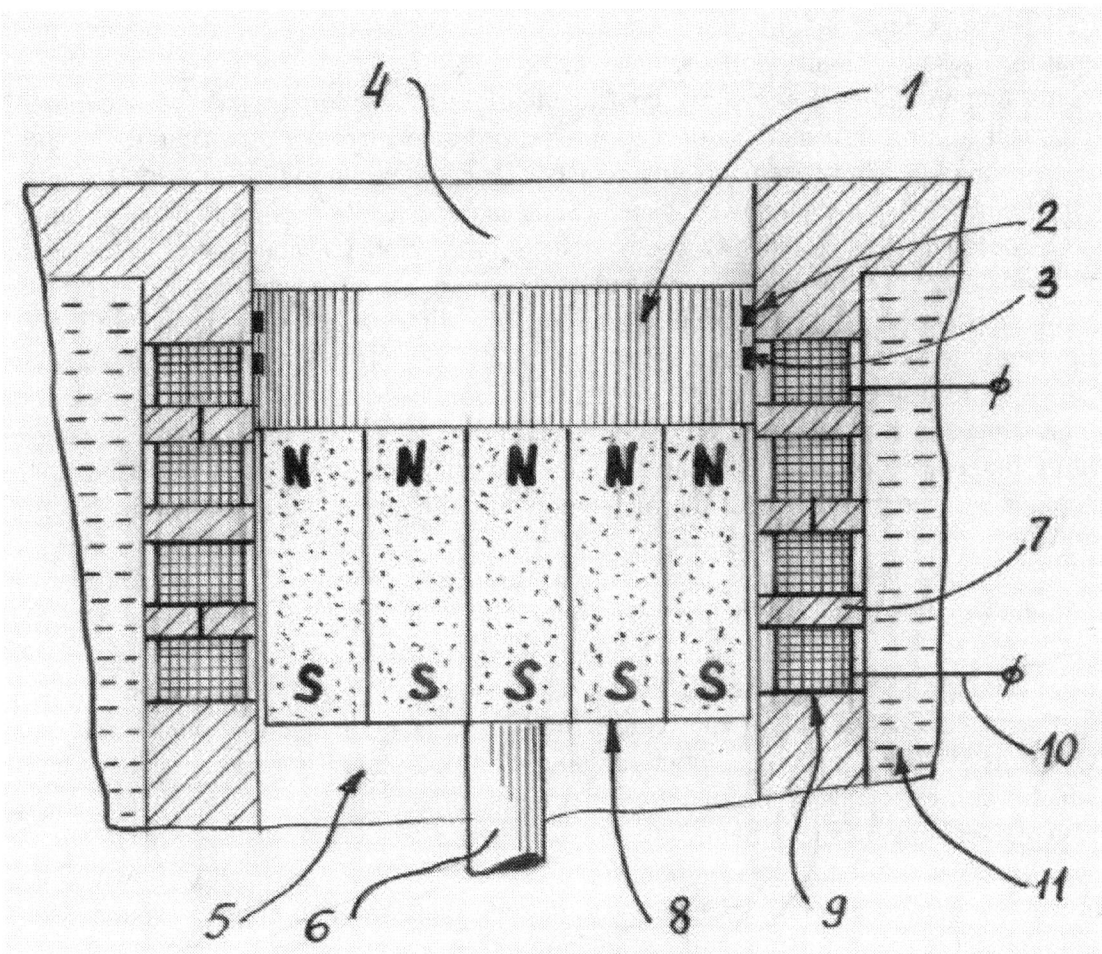

1 CLAIM, 2 DRAWING FIGURES

COMBINED ENGINE

BACKGROUND OF THE INVENTION

The present invention relates to engineering advantageously to the combustion and gasoline engines having the crankshaft which turn the thermal energy into mechanical, i.e the progressive movement of piston changes to rotating movement of crankshaft) and forming additionally a new kind energy-electricity energy in period of this changing.

At present time the changing one kind of energy to another in a engine mechanisms realizes differently. In U.S Patent 4,876,992 the progressive movement of a movable element of piston changes a rotating movement of crankshaft. In this occasion changes only the kind of mechanical movement-progressive movement changes into rotating movement and this show a general defect of this device because does not make any additional energy for this period of changing.

This defects is removed in construction (C.E. Matson "30 instruction units in basic electricity "- McKnight Publishing Company, Illinois, 1961, pp.59-61) where in a gasoline engine, consisting from a fundamental crankshaft engine and generator device installed in this engine autonomously.

In this engine changes a thermal energy to mechanical and then to electricity energy.

The general defect of this device shows that there is a functional dependence of this changing: a gasoline engine works as a generator works.

This engine shows a very low coefficient of useful action (c.u.a) for account of a mechanical loss on transmission of a generator and also increasing a material-capacity of power installation. This system advantageously characterizes automobile transport.

In another well-known a power system (Science and Invention Encyclopedia, volume 1 -Connecticut,1989, pp.52-54) these defects are removed and there is a separate functional design of mechanisms: a general engine auxiliary engine-generator.

This system advantageously characterizes maritime ships (cargo and military of designation) and sometimes characterizes auto-transport system—separate car has a general engine with to fastened auxiliary engine generator for fulfilling the different works. In this scheme a general engine changes the thermal into the mechanical energy with transfer of a rotating movement to the transmission. And an auxiliary engine-generator consisting from the engine and generator changes thermal energy into the mechanical and then the electricity energy.

The general defect of this system is a very low coefficient of useful action (c.u.a) and very high a material-capacity of power installation because a general engine and auxiliary engine-generator are separated.

The attempts to increase the coefficient of useful action (c.u.a) of engine are showed in U.S. Patent 4,342,920. The engine converting gas expanding energy sources into useful other forms of energy such as inducing an alternating current in a load-connected electrical conductor coil by means of surrounding of cylinder of said engine and holding a piston by means of forming a magnetic field.

In this engine utilizing gravitational force of piston in the movement of down. But this device has the different objectives and need in additional installation as the compressor for the movement of the piston to the upper position.

These defects are removed at present invention which appointed in advance for the increasing of coefficient of useful action (c.u.a) engine and decreasing of material-capacity of the system in whole by means that in a crankshaft engine, beside of changing of the thermal energy into the mechanical at this engine additionally forms the electricity energy at period of working process and this changing makes into account of the special installation at this engine.

SUMMARY OF THE INVENTION

At present the invention appointed in advance for increasing of a coefficient of useful action (c.u.a) of engine and decreasing of material-capacity of power system by means of forming additionally at this engine generator and receiving from this device the electricity energy besides general energy forming from the crankshaft energy, thermal energy, and thes facts considerably improve the technic-economical indexes of any engine (reducing the specific expenditure of fuel, rising of effectiveness of power system and etc.).

The fundamental prerogative of present invention proves a high effectiveness of coefficient useful action (c.u.a) by compare with well-known engines (value of c.u.a composes 0.3).

The other important goal of the present invention proves reducing of material-capacity well-known engines on the whole complex power installations (engine, transmission, system of electricity, etc.) prerogatively in, aritime ships and submarine particularly where this factor is more important.

This and further objects of this present invention will be described in the following description and claims, and are illustrated in the accompanying drawings, which, by way of illustration, show a preffered embodiment of the present invention and the principles thereof, and what is now considered to be the best mode in which to apply these principles.

Other and different embodiments of the invention emboding the same or equivalent principles may be used and structural changes may be made as desired by those skilled in the art without departing from the invention.

BRIEF DESCRIPTION OF THE DRAWINGS

Figure 1 depicts a cutting full section, front view of a combined engine of the invention;
Figure 2 depicts removed section of a combined engine of Figure 1.

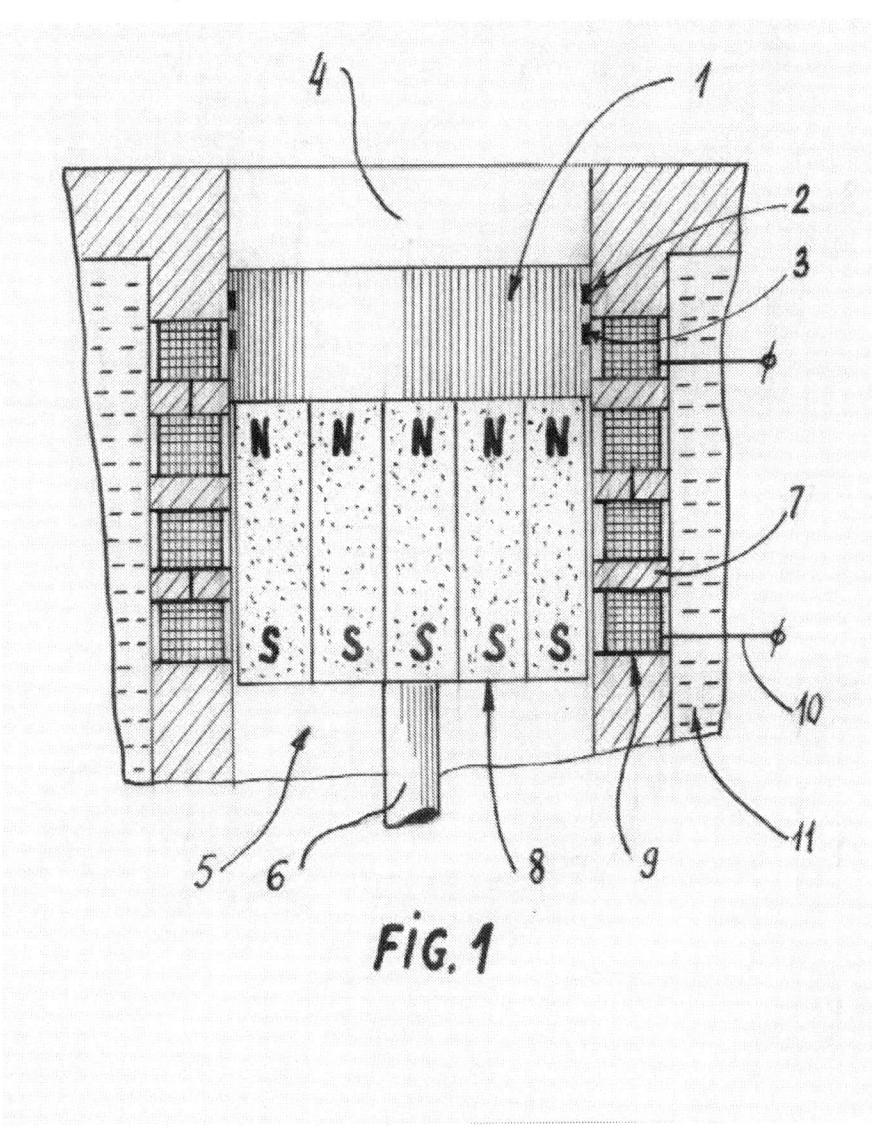

FIG. 1

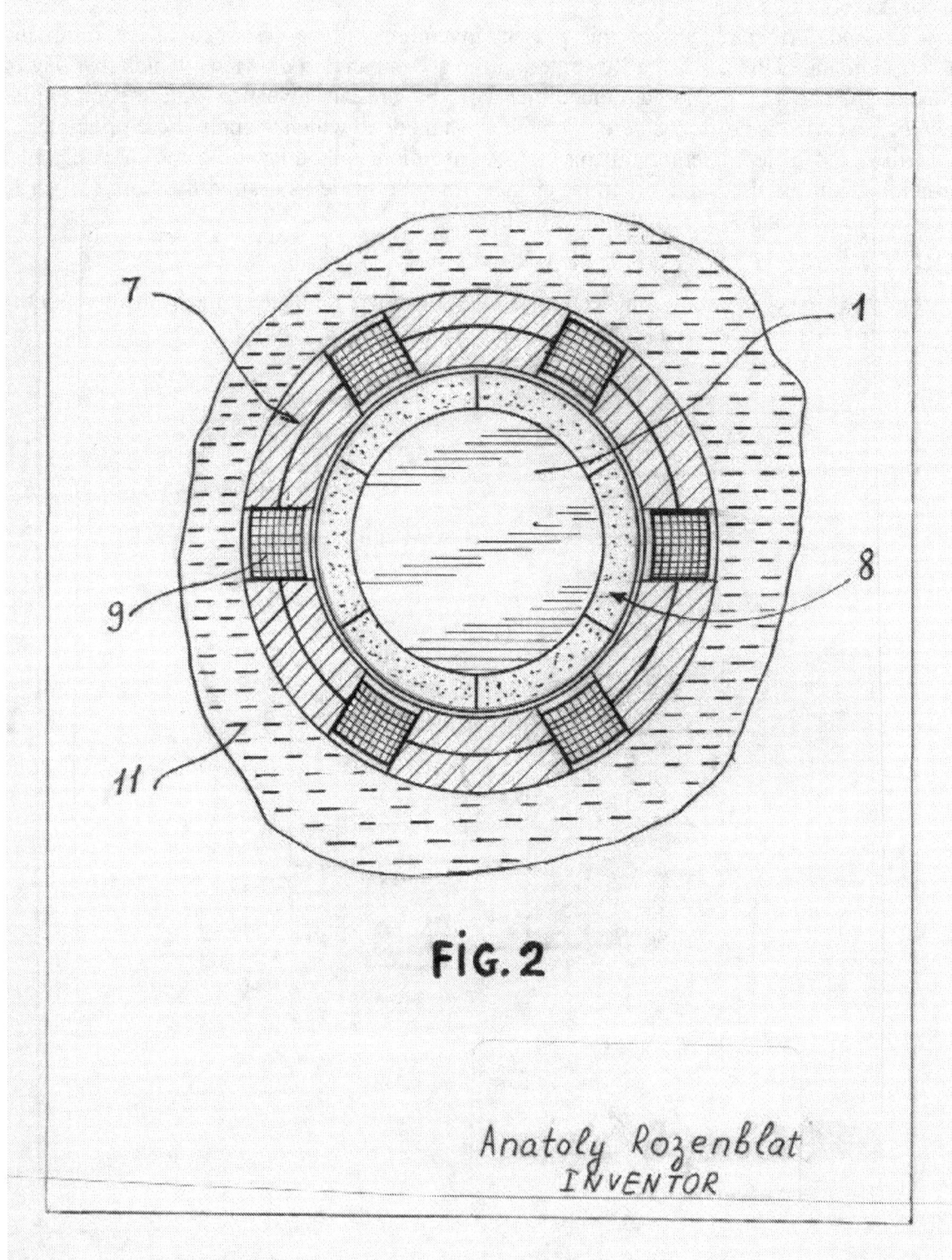

FiG. 2

Anatoly Rozenblat
INVENTOR

DETAILED DESCRIPTION OF THE DRAWINGS

Referring now to the figures and particularly to Figure 1 and 2 we see a fundamental contents of the present invention.

The combined engine consists mainly from the thermal crankshaft engine which is presented in the form of the basic parts: a movable element in a form of piston 1 constantly removaling in the progressive direction, on which are fulfilled the ringe-shaped hollows at periphery and installed the compression rings 2 and oil rings 3 to be prevent falling of combustible mixture from the combustion chamber 4 to crankcase 5, and joined with crankshaft 6 for transfering of movement to a executive mechanism, for example to propeller (on drawing does not shown), and a motionless element embracing to said piston 1 which is fulfilled in form of a cylinder 7 (or bung of cylinder).

In accordance with the present invention to in a combined engine gets also additionally is formed engine-generator which consists from the following elements:

on the periphery of a piston 1, on his length of circle, are installed permanent or other magnets 8 and they are fixed relatively of the cylinder 7 as shown schematically in Figure 1 and 2.

The permanent magnets 8 bcfore covered and fixed by rather thin thermal layer which is made, for instance, from titanium alloy and all this installed in underneath part of piston 1. A winding of excitation 9 is installed in a hollow of said of cylinder 7 incontactly with the piston 1 the following way:In a cylinder 7 fulfils the hollows and at the length of circle this cylinder 7, as shown in Figure 2, the separate coils 9 are arranged which are joined one with another in series on perimeter to said of cylinder 7 with the exit to both ends of wire 10 to consumer of energy (not shown on the drawings).

On the side spaces of this separate coils 9, one situated to the cylinder 7, and other to a water system 11 of engine cover before and fixed also by the thin thermal layer which as the sample is made from the titanium alloy and installed they later in the hollows of this cylinder 7. Looking at the Figure 1 and 2, we see that the permanent magnets 8 and the separate coils 9 (the winding of excitation) are installed with the space.

• In during of working process in a crankshaft engine a movable element in a form of piston 1, on which are fixed the permanent magnets, constantly removaling in the progressive direction along of cylinder 7, in which are installed the winding of excitation 9 (the separate coils 9) intersect the magnetic field of these separate coils 9 (the winding of excitation 9).

• In during fulfilling of the thermodynamical process to a crankshaft engine in a piston 1 with the permanent magnets 3 (function of solenoid perform the piston 1) forming the magnetic field which on Farady's law, in a winding of excitation 9 (the separate coils 9) the last performs the function of inductance-coil which induces a electromotive force (e.m.f) and current.

It should now be apparent that the present invention provides in increasing coefficient of useful action (c.u.a) in combined engine and decreases a material-capacity of system in the whole by means that in a engine besides of changing thermal energy into mechanical forms additionally the electricity energy at period of the working process for account that in special installation of this engine there is a generator, as the new element.

Thus, while a preferred embodiment of the invention has been illustrated herein it is to be understood that changes and variations may be made by those skilled in the art without departing from the spirit and scope of appending claims.

I CLAIM:

1. A combined engine for converting energy of thermodynamical process of this crankshaft phasing engine into mechanical energy to one side such as transferring motion to mechanism of transmission having one or more movable axially piston whereby connected rod and cylinder of said engine *comprising to the other side parallel forming additionally electric energy in a special generator device which at least has the permanent or other magnets rigidly mounted on said movable piston in the progressive direction and which are fixed on its periphery, forming the closed surface with said cylinder, and finally the magnets are installed in lower part of said piston on the side of crankcase space so, that elements of the coils firmly connected in series circuit and installed inside of said cylindrical surface and generatrix element of said cylinder to the special made hollows in a form of said coil and their contacting surfaces with the said magnets are arranged with minimal space relatively to said cylinder and piston, and all this system forms the current for one working process of engine.*

The advantages of this invention is introduced separately below of this material.

DESCRIPTION OF ADVANTAGES OF THE INVENTION:

« COMBINED ENGINE «

USEFULNESS:

The fundamental prerogative of present device proves a high effectiveness of coefficient useful action (c.u.a) by compare with the well-known engine (the value of c.u.a which is about of 0.30).

The other important objective of this device proves by reducing of material-capacity of complex power installation (engine, transmission, system of electricity) mainly for the maritime ships and submarine particularly where this factor is more important.

NOVELTY:

In this device the piston has the permanent or other magnets which are rigidly mounted and fixed on periphery of lower part of said piston.

And cylinder of this engine has the elements of coils including one or more separate coils which are connected in a rigid unitary curcuit with other elements in series circuit and the last installed inside of said cylinder on its perimeter along of his cylindrical surface and generatrix element of said cylinder to the special made hollows in a form of said coil and their contacting surfaces with the said magnets are arranged with the minimal space relatively to said cylinder.

ORIGINALITY:

In the process of committing a full working cycle in a crankshaft engine, the piston with the fixed magnets (they fulfill a function of solenoid) produce a magnetic field which on Farady's law in a winding of excitation is installed in a hollow to said of cylinder with the coils (they fulfill a function of inductance—coil) induce a electromotive force (e.m.f) and additionally current.

PROFITABLE:

In this device provides by means of extracting mainly the diesel-generators from the engine rooms, for instance in submarines or ships that considerably decrease the material capacity of complex power marine system and accordingly increases the coefficient of useful action of engine and improve the manoeuvre-tactical characteristics of vessel in whole.

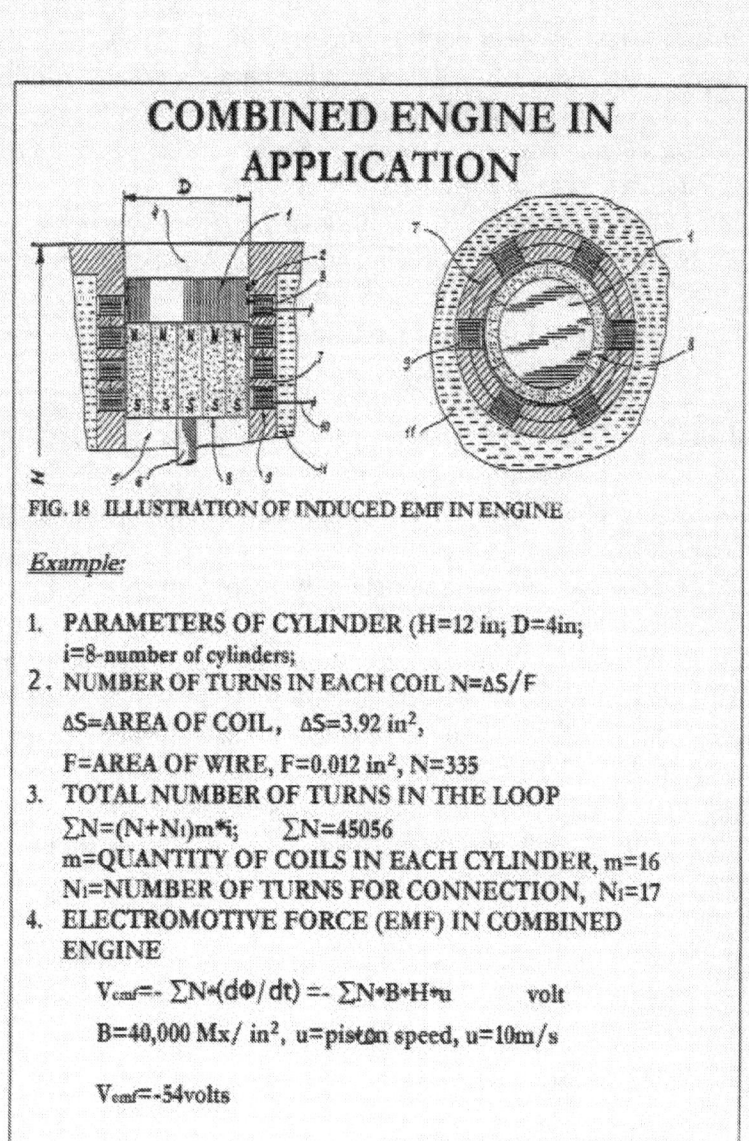

COMBINED ENGINE IN APPLICATION

FIG. 18   ILLUSTRATION OF INDUCED EMF IN ENGINE

Example:

1. PARAMETERS OF CYLINDER (H=12 in; D=4in;
   i=8-number of cylinders;
2. NUMBER OF TURNS IN EACH COIL N=ΔS/F

   ΔS=AREA OF COIL, ΔS=3.92 in^2,

   F=AREA OF WIRE, F=0.012 in^2, N=335
3. TOTAL NUMBER OF TURNS IN THE LOOP
   ΣN=(N+N$_1$)m*i; ΣN=45056
   m=QUANTITY OF COILS IN EACH CYLINDER, m=16
   N$_1$=NUMBER OF TURNS FOR CONNECTION, N$_1$=17
4. ELECTROMOTIVE FORCE (EMF) IN COMBINED
   ENGINE

   $V_{emf} = -\Sigma N*(d\Phi/dt) = -\Sigma N*B*H*u$ volt

   B=40,000 Mx/ in^2, u=piston speed, u=10m/s

   $V_{emf} = -54$volts

13. Rozenblat's combined vehicle

BACKGROUNG OF THE INVENTION

1. Field of the invention

The present invention relates advantageously to automobile industry and other transportation vehicles which use the gasoline and combustion engines. And this invention also could be used also in chemistry industry.

2. Prior Art

At present any industrial source or mechanism having any engine makes a lot of combustion gases which are very toxic for the environment (Standard Handbook for Mechanical Engineers.- McGraw-Hill book Company,1967,9.-104).

In accordance with the data of author (Fundamentals of air pollution by Samuel J. Williamson,1973, p.102.- Addison-Wesley Publishing Company): "*... that rapidly mounting pace of industry might release sufficient carbon dioxide CO_2 to cause climate changes by affecting the surface temperature of the earth....*"

Average carbon dioxide emissions to the atmosphere from fossil fuel combustion to be increasing as more use of automobile for transportation. As shown by the author in Table 4,118 (Air pollution and conservation by Jan Rosvall,1988- Elsever, Amsterdam) carbon dioxide from combustion of fossil fuels is increasing by about of 2 ppmv/year.

SUMMARY OF THE INVENTION

The principal object of the present invention is to provide a device for reducing of combustion gases from the different sources (engine, plant, etc.) to the environment space. It also is an object of the present invention to provide such a device and method which will be useful for the society.

The foregoing objects can be accomplished by providing an exhaust pipe which is fulfilled in view of vessel conformable to a combustion engine. At least on end of exhaust vessel connected with the source forming the combustion gases and other part of said vessel have not joined with the environment space. And besides the vessel is made, for instance, in view of emptiness rectangular box where the lower part of said vessel is fulfilled with incline. The vessel at this case are closed in both sides by two covers which are joined firmly with the coils and pistons moving axially across of said vessel.

A device is installed so that in inner space of said vessel inputs incoherently the spherical emptiness balls which have the spring-and—ball valves (non-returned) contacting one with another and also two covers for the pistons. The above-named balls absorb the combustion gases directly from the inner space of the vessel and then storage these said gases inside of the balls with the next following operation—unloading to the special installation for reproduction of these gases and use in chemistry industry.

BRIEF DESCRIPTION OF THE DRAWINGS

A combined vehicle embodying features of our invention is illustrated in the accompanying drawing, forming part of this application, in which:

FIG.1—is a front view of combined vehicle accordingly with the present invention;
FIG.2—is a fragmentary, top view of the bottom end portion of the device of FIG.1;

FIG.3—is a fragmentary, right side view of combined vehicle, in section A-A of the FIG.1;

FIG.4—is a broken—out section top view of combined vehicle;

FIG.5—is a front view of combined vehicle with the special installation (unloading process) accordingly with the present invention.

DETAILED DESCRIPTION

FIGURES 1,2,3,4 and 5 show the simple drawings which are suggested by the author for realization of this proposal for commercialization in industry.

Automobile 1 has the engine 2 with the exhaust pipe 3 which is connected with the special vessel 4 having spherical balls 5 and installations 6 on both sides of this vessel 4.

The installation 6 has movable elements 8 with the coils 9 and cover 10 which fixed firmly the back side of this vessel 4 and has also the micro-switch 11. The installation 7 also has the movable elements 12 with coils 13 and cover 14 which firmly fixed in the front of the vessel 4 and also has the micro-switch 15.

The floor inside of car additionally is covered by the thermal insulation 16. The spherical balls 5 have the spring-and- ball-valve (non-returned) 17 which firmly fixed in the body of balls 5.

- In the process of working engine 2 the combustion gases come into the special vessel 4 which is closed in both sides firmly. At this period the pressure in vessel 4 increases and in the next time the spring-and-ball valve 17 (not-returned) opens in any random spherical balls 5.

In fact that the combustion gases 18 come into the balls 5 and naturally the pressure in closed vessel 4 at this period falls down, i.e this process makes in each spherical balls 5 as was indicated above occasionally at this working process of engine.

The material and thickness of each spherical ball 5 must be calculated for definite pressure and besides the quantity of balls 5 which are input at this vessel 4 also must be calculated and necessary to admit that all this depends from period of the working process of engine and total pressure at this period.

If the pressure in during of working process in engine increases then the movable elements 8 and 12 begin to move axially with the coils 9 and 13 so that micro-switch 11 and 15 make the "switch off" process of engine or gives the emergency sound to the driver of this vehicle that the pressure of the exhausted gases is overcharge and necessary to exchange the balls 5 on the special installation.

Author thinks that also possible to install and fix the special valve 21 for the safety pressure in the vessel 4(emergency version).

On the next step the driver should moves to the gasoline station, as shown in Figure 5, that to exchange the balls 5 which are already filled by the combustion gases. In this period the old balls 5 (filled with the gases) fall down to the special underground installation 20, for instance and the driver receives the new balls 19.

I CLAIM:

1. Combined vehicle comprising:
   a. an exhaust pipe which is fulfilled in view of vessel conformable to a combustion engine.

    b. at least one end of exhaust vessel connected with source, forming of combustion gases and the other parts of said vessel have not been joined with the environment space.

    c. A vessel is made, for a instance, in view of emptiness rectangular box where the lower part of said vessel is fulfilled with the incline.

    d. A vessel closed in both sides by two covers which are joined firmly with the coils and pistons moving axially along of said vessel.

2. A device for the reducing of combustion gases in environment space comprising:

    a. at inner space of said vessel inputs incoherently the spherical emptiness balls which have the spring-and—alls valves (not-returned) contacting one with another and two cover for the pistons.

    b. said balls absorb the combustion gases directly from the inner the space of vessel and the storage of said gas inside of these balls with the next operation of unloading tot he special installation for utilization on the chemistry industry.

COMBINED AUTOMOBILE AND METHOD OF REDUSING COMBUSTION GASES

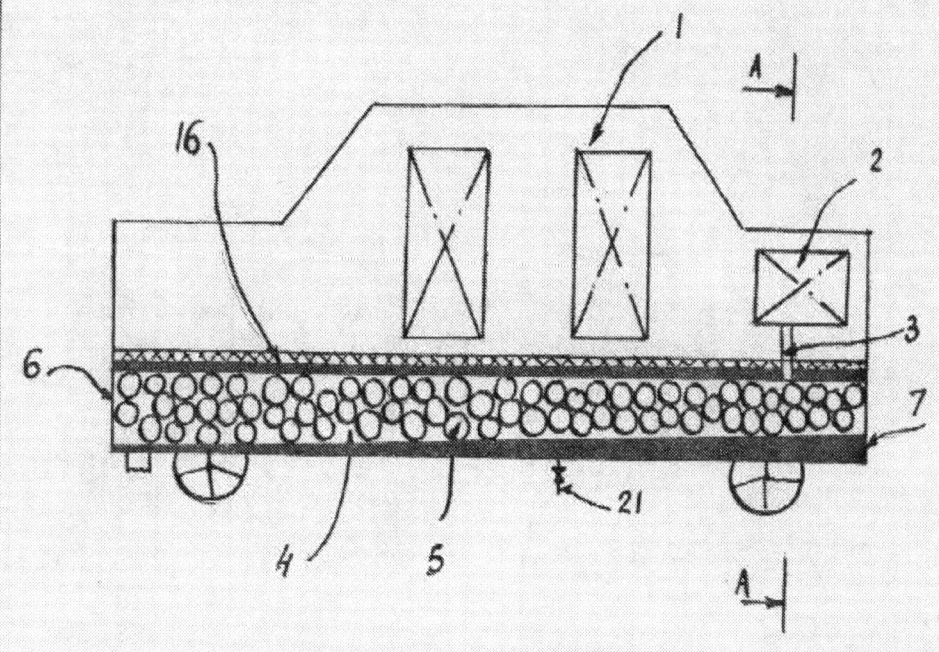

FIG.1

INVENTOR

ANATOLY I.ROZENBLAT

COMBINED AUTOMOBILE AND METHOD OF REDUCING COMBUSTION GASES

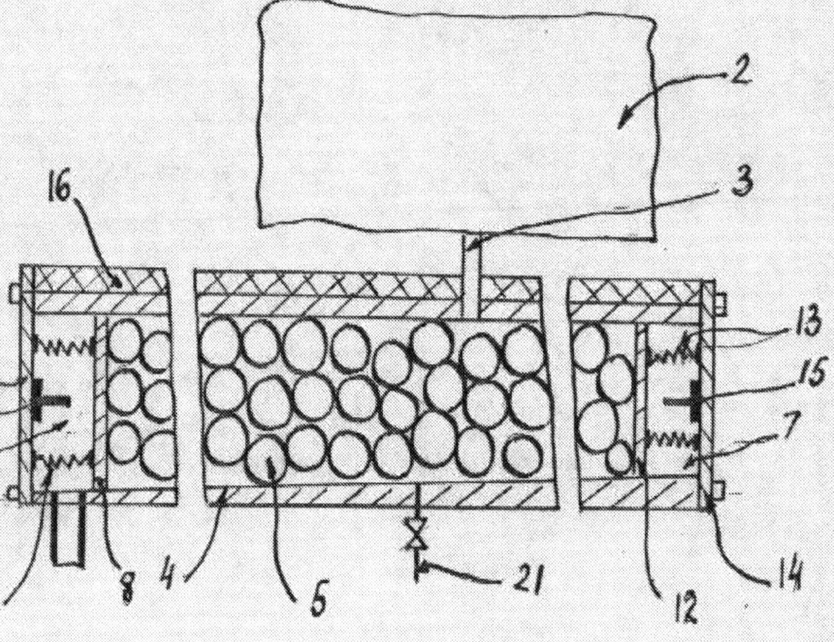

FIG.2

INVENTOR

ANATOLY I. ROZENBLAT

COMBINED AUTOMOBILE AND METHOD OF REDUCING COMBUSTION GASES

A-A

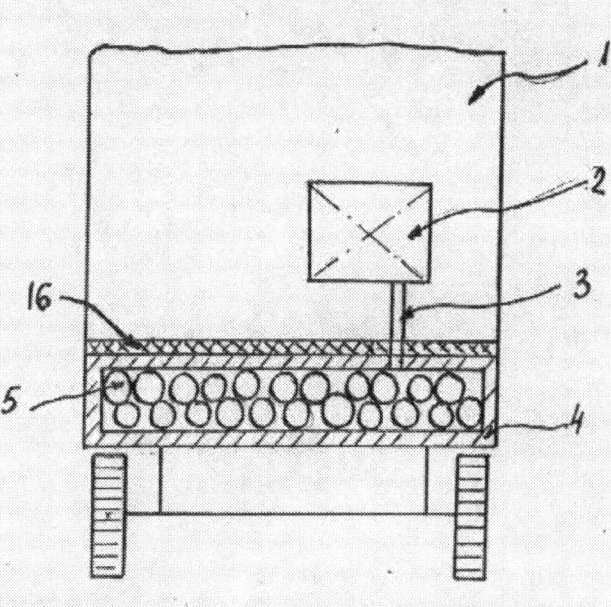

FIG.3

INVENTOR

ANATOLY I. ROZENBLAT

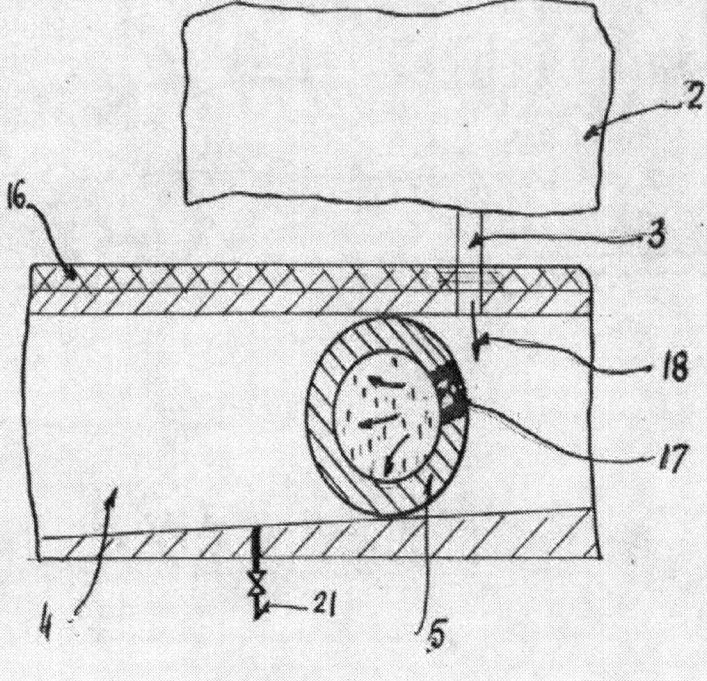

COMBINED AUTOMOBILE AND METHOD OF REDUCING COMBUSTION GASES

FIG. 4

INVENTOR

ANATOLY I.ROZENBLAT

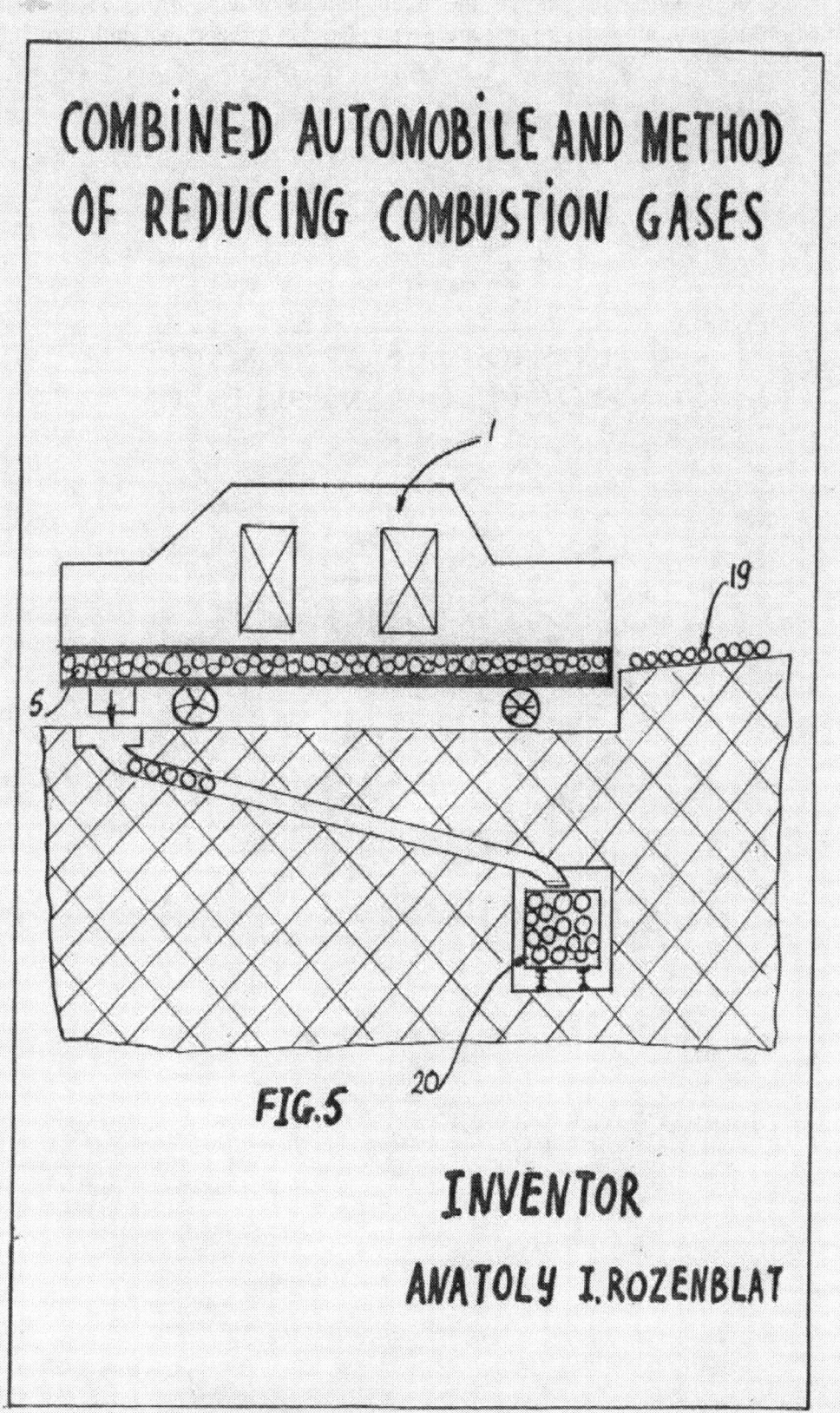

COMBINED AUTOMOBILE AND METHOD OF REDUCING COMBUSTION GASES

FIG.5

INVENTOR

ANATOLY I. ROZENBLAT

On the next pages are shown the main factors which influence on rising of the combustion gases to atmosphere and are given also some recommendation of improving of this problem.

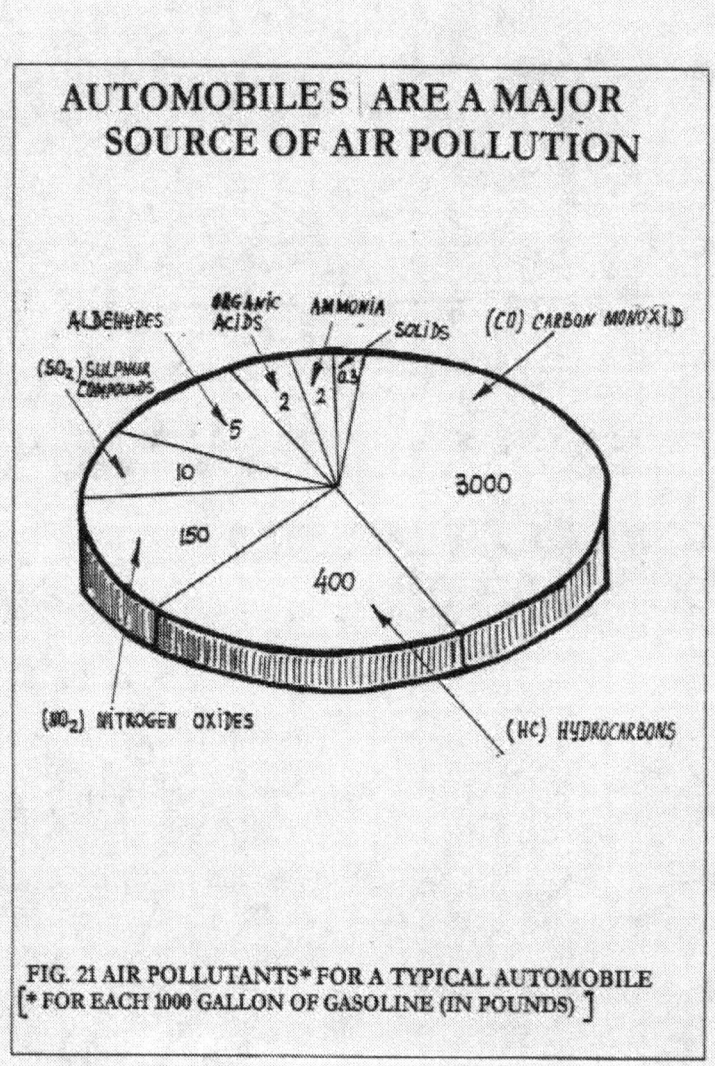

FIG. 21 AIR POLLUTANTS* FOR A TYPICAL AUTOMOBILE
[* FOR EACH 1000 GALLON OF GASOLINE (IN POUNDS)]

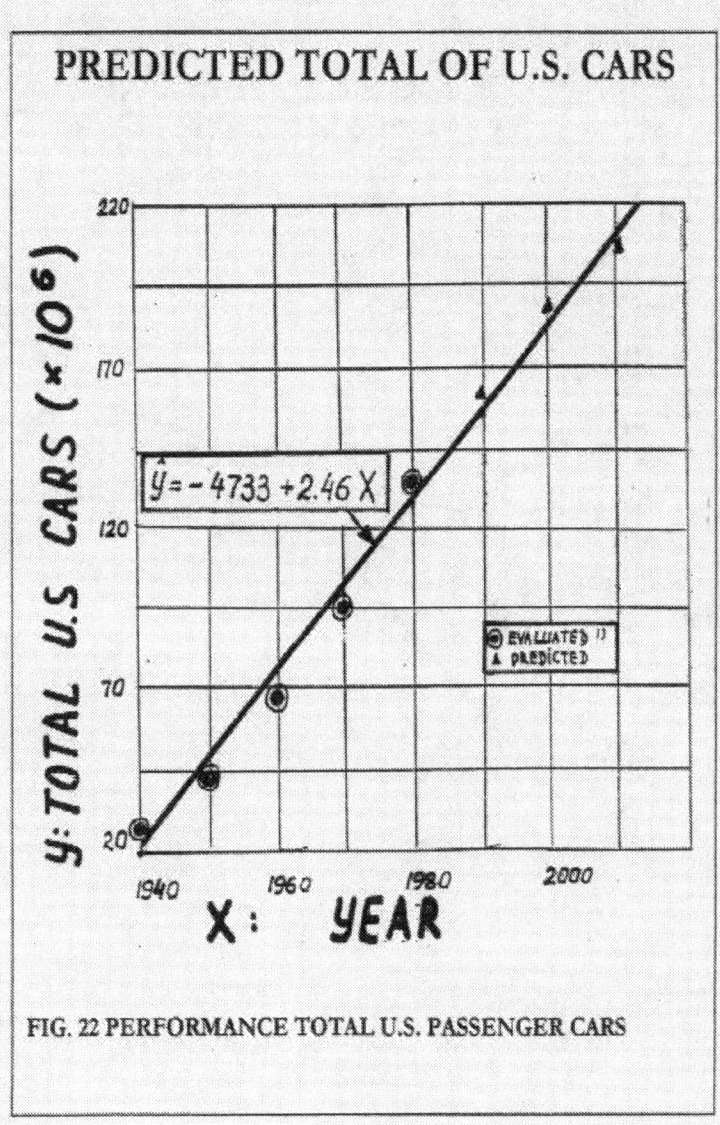

FIG. 22 PERFORMANCE TOTAL U.S. PASSENGER CARS

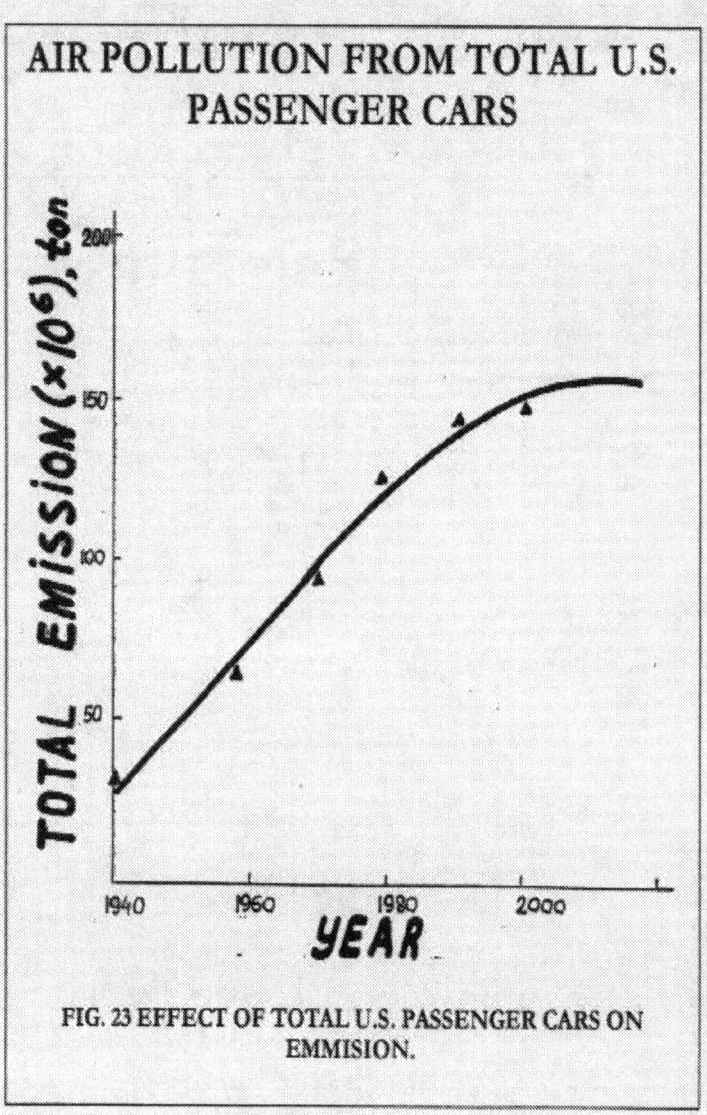

FIG. 23 EFFECT OF TOTAL U.S. PASSENGER CARS ON EMMISION.

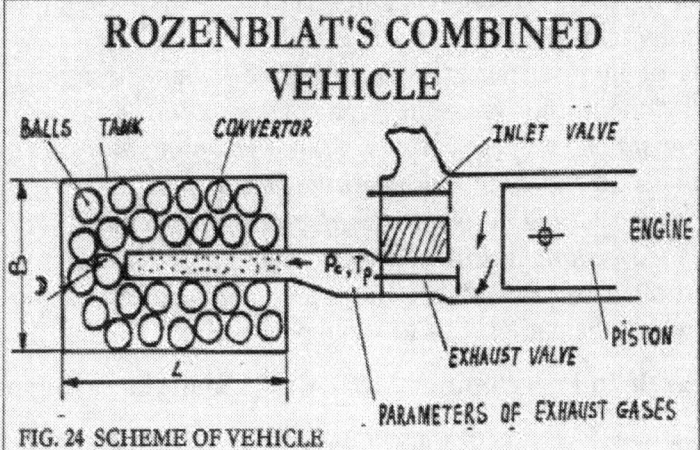

FIG. 24 SCHEME OF VEHICLE

Example:

1. EXHAUST PORT EXIT TEMPERATURE T_P=800K(1400R)
2. EXHAUST MASS FLOW RATE m_e=30g/s
3. CYLINDER PRESSURE P_C=140kPa
4. PARAMETERS OF TANK (L=36in; B=24in; H=8in)
5. VOLUME ONE BALL, V_1=0.524*D^3, V_1=14 in^3 (R=1.5in, r=1in);
6. QUANTITY OF BALLS n=494
7. TOTAL EMISSION IN TANK, M= $m_e \cdot \Delta$, M=1900lb
8. PRESSURE IN TANK P=(M*R*T_P)/V
   V=VOLUME OF TANK, V=61912in^3.
   R=GAS CONSTANT, R=$\bar{R}$/K; $\bar{R}$=1545ft-lb/lb mole R
   K=28.01 lb/lb mole (FOR CO); R=55.1ft-lbf/lbmR
   SO, P=2750lbf/in^2
9. PRESURE P_1 IN EACH BALL P_1=5.6lbf/in^2

10. AREA OF ONE BAL, S=4π(R^2-r^2), S=15.7 in^2
11. FORCE ON THE WALLS OF EACH BALL F_C=P_1*S=88lb(39kg)

14. Some advanced technology for the metal processes

A. Introduction

In metal-cutting processes of a work-piece the problem of decreasing of the cost production and its precision is the main objective for any manufacturing plant.

Many researches work at this direction for a long time that to design the new different cutting tools and elaborate the methods which could improve thew metal—cutting processes in whole (*Nakayma K.,1962. " A study on chip-breaker".- Bulletin of the Japanese Society of Mechanical Engineering, vol.5, No.17, pp.142-150*).

Some attempts of improvement the cutting process and designing of the new cutting tools are shown by the author in some works (*Rozenblat, A.I., 1996 "Analysis of chip formation in cutting stainless steels with circular segmental saws".-Proceedings of the 26th Israel Conference on Mechanical Engineering, Haifa, pp.138-140*) and (*Rozenblat, A.I., 1998 "Rozenblat's new cutting tools for manufacturing processes ".- Proceeding of the 27th Israel Conference on Mechanical Engineering, Haifa, pp.427-432*).

But despite on all these advantages there are many unsettled questions in the metal-cutting processes which the author tries to solve positively.

B. The basic research and development activities in manufacturing process.

Considering the new methods of machining of work-piece and also the different cutting tools for their realizing we must admit that above-named advanced technology mainly recommends for such important tasks as:

1. *Increasing of precision of work-piece into account of decreasing of its temperature deformation in cutting processes;*
2. *Increasing of tool life;*
3. *Improving of production effectiveness into account of decreasing the auxiliary time on maintenance and adjustment operations for cutting tools and processes;*
4. *Improving of break- chip process in cutting period advantageously for the stainless and high-resistance steels;*
5. *Improving of machinists health and manufacturing processes into account of using the pneumatic transportation for removal of chip and dust from the cutting zone of machine tools.*

C. New applications

Feasibility of suggested innovative applications related to Manufacturing Engineering and more concretely to the machining processes with the using of single—point cutting tool and pneumatic transportation which have been investigated with a big success in industry.

14.1 METAL-CUTTING PROCESS WITH VARIABLE DEPTH OF CUT

In well-known turning and milling processes (*Pollack, H.W. "Tool design".- Prentice Hall, Englewood Cliffs, N.J 1988, pp.342-344*) the depth of cut is the value constant.

However, such scheme of cutting increases the total temperature of process and heating of work-piece considerably. As the result of this high temperature, the form of work-piece distorts and tool life decreases accordingly because the temperature deformations presence at this process.

The author thinks that in the future the new suggested method of cutting could decrease the temperature of work-piece and improves its precision in whole.

There are two major reasons for increasing of accuracy and precision of work-piece in metal-cutting processes:

- The first- has been an ever-increasing of demand for the form of detail and dimensional accuracy and their repeatability from a manufactured product quality perspective;
- The other—has been due to increasing automation of the equipment with the commensurate of operator input and skill to overcome potential errors in the machine tool.

This objective can be achieved by using of the metal-cutting process with variable depth of cut which guarantees the precision of cutting the work-piece and increases the tool life into account of decreasing of its temperature deformations.

Metal-cutting process of cylindrical work-piece with variable depth of cut has been developed by the author of this material. The most important advantages of this method is the increasing of precision of work-piece in process of its cutting and also the increasing of tool life into account of decreasing of its temperature deformations.

The main point of this method consists in that the total allowance given on the cutting of work-piece divide on the some intermediate allowance. And then each intermediate allowance divide exactly on two pass of cutting.

After of these primary procedures the work-piece at the first pass of cutting process displace on some angle α which is equal:

$\text{Tan } \alpha = A_{max} / L$

Where,

A_{max} = maximum value of intermediate allowance on each pass;

L = length of cutting of work-piece.

At the first pass in cutting process the depth of cut for this work-piece is value variable: the value is equal maximum at the beginning process and the value is equal zero at the end of cutting for this intermediate allowance.

After of finishing the first pass of cutting the work-piece arranges at primary position with value of angles which is equal zero ($\alpha = 0°$) and makes the second pass of this intermediate allowance.

The peculiarity of the second pass that the value at the beginning process is equal zero and the end process is equal maximum.

The metal-cutting process schematically is shown in Figure 1 in the different variants:

- Figure A shows the first pass of cutting of this work-piece;
- Figure B shows the second pass of cutting of this work-piece;
- Figure C shows the cross-sectional view of A-A on Figure A;
- And Figure D shows the value of square of chip (s) for each pass in dependency of the length of cut (L) of this work-piece.

- At the first pass (Figure A) work-piece 1, having the central axis 2, fixes in the jaw chuck (not shown) and centers 3. Then the back center or tail spindle of tail-stock (not shown) displaces on the angles 4 securing the position of work-piece at the central axis line 5. And after of this operation, the cutting tool 6 fixes and the work-piece gives the rotation 7 and feed 8 to the cutting tool 6.

At the first pass from the outer surface of work-piece 1 cut the variable depth of cut 9 so that it value is equal maximum at the beginning process and its value is equal zero at the end of process cutting. So, after of finishing of the first pass the work-piece 1 has the outer surface in view of cone, as shown in Figure B.

-At the making the second pass the work-piece 1 arranges again to the central axis of machine tool in the position 2, i.e the axis of machine tool and feed of cutting tool is located parallel. The cutting tool 6 at the second pass (see Figure B) gives the feed 8 and begins the machining process for the second variable allowance 11 from the outer surface of the work-piece.

The value of this allowance at the beginning process is minimum and to the end of machining process of the work-piece the value is maximum. In both case, as shown in G

Figures C and D, the square of cut changes and has the variable character.

In view of the fact that the length of contact of the leaving chip on the front its surface of the cutting tool has the variable value and we can do conclusion that the tool life at this case increases considerably and also the total quantity of heating in cutting zone decreases in whole.

So, for this reason the precision of the cylindrical work-piece 12 and tool life 6 improves into account of decreasing of the temperature deformations in cutting zone.

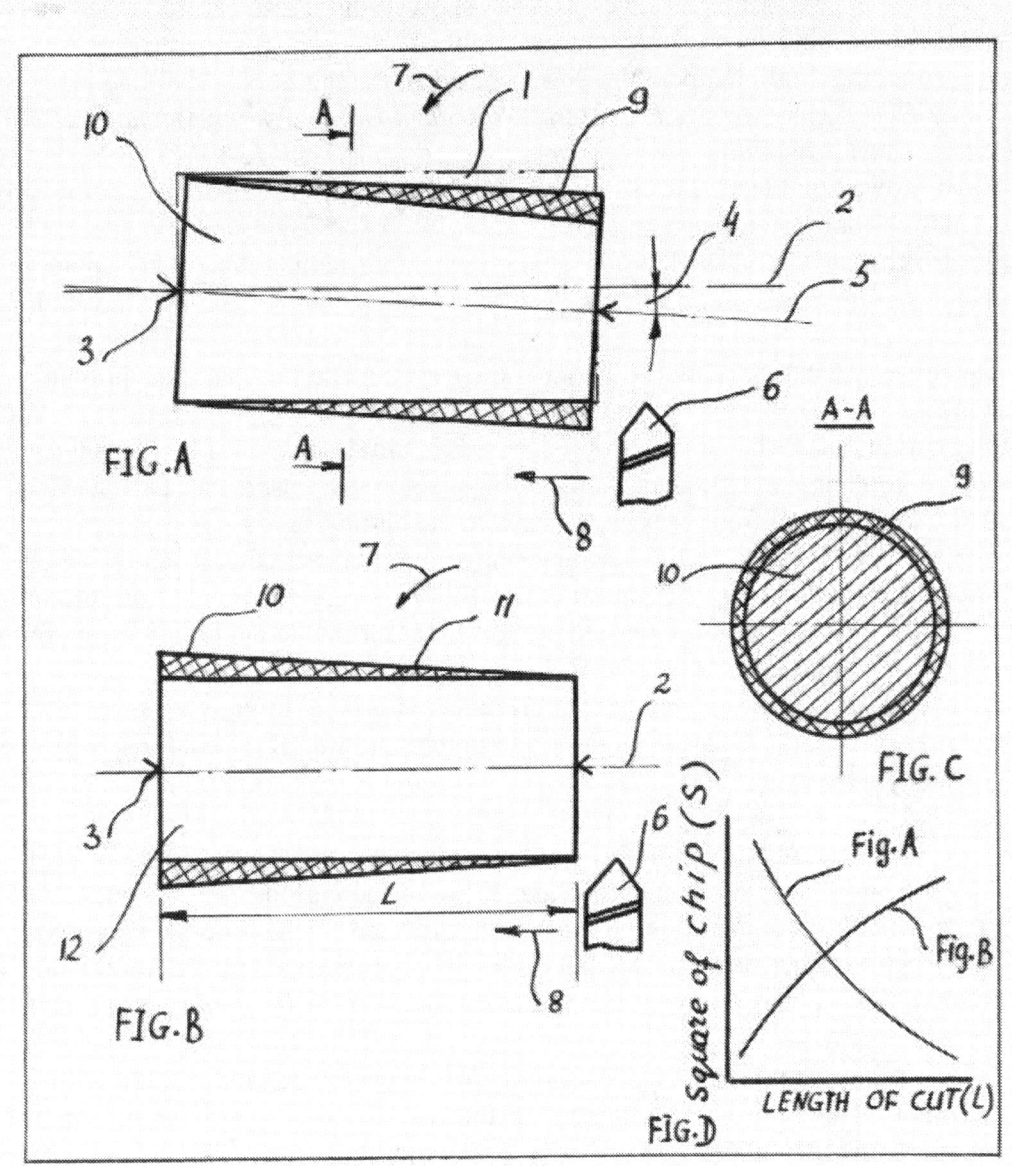

FIGURE 1 METAL-CUTTING PROCESS OF CYLINDRICAL WORKPIECE
WITH VARIABLE DEPTH OF CUT

14.2 HEAT-REMOVAL CUTTING INSERT

The different cutting carbide inserts have been described by ISCAR Metals Inc.(*ISCAR Metals Inc. "21ˢᵗ Century Catalog", 1988, pp.140-141*).

However, the experimental investigations showed that these carbide inserts as and other do not have the high overall strength and splitting strength for the reason of presence in cutting zone the high temperature. And besides the authors, Avallone and Baumeiter (*Avallone E.A, Baumeister III T., "Mark's Standard Handbook for Mechanical Engineers. 9 th Edition._ McGraw-Hill Book Company, New York, 1987, pp. 13-52, 13.55*),

Have showed that hardness of tool insert decreases with the rising of cutting temperature for all cutting-tool materials and inserts.

So, for the carbide insert the decreasing of heating temperature is the main goal for all metal-cutting processes and particularly for the cutting of the stainless and high-resistance steels.

This objective can be achieved better if the carbide insert to make the porosity, as shown in Figure 2.

The carbide insert has the cutting edges 1,2,3 and 4 so that on the perimeter they form the closed contour. And besides these cutting edges, the insert has two surfaces 5 and 6 which could use as the front or back surfaces in the cutting process.

The surfaces 5 and 6 on the perimeter have the inclined cutting edges which are made with the different front angles, for example with the negative angles 7 on all cutting edges 1,2,3 and 4 of this insert. In the middle part of this insert made the hole 8 for its fixing on the holder of this cutting tool (not shown).

As above—indicated, the carbide insert has the porosity in view of micro-holes 9 which could be made in the process of high-temperature sintering of carbide insert by the method of powder metallurgy.

- In process of cutting from the work-piece the total quantity of heat distributes between insert, chip and tool. As the carbide insert is made in view of porosity and more detail with thorough micro-holes 9, as shown on this construction, we see that on the laws of thermal convection the big quantity of heat removes to the environment from the cutting zone and carbide insert.

- So, at this case the heating of work-piece and carbide insert will be minimal and this fact promotes for the increasing of precision of work-piece in cutting process and also the tool life improves in whole into account of decreasing of temperature deformation.

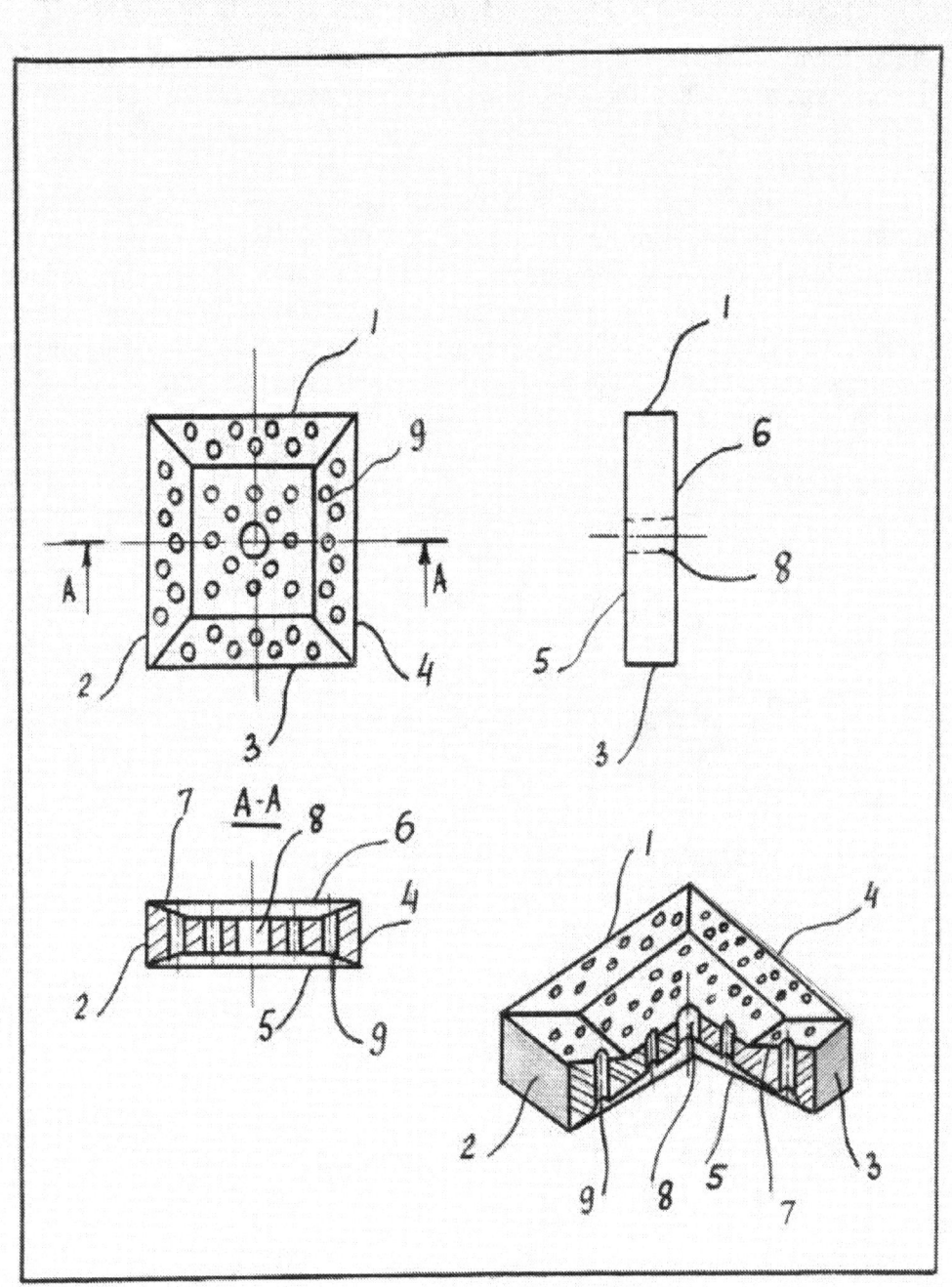

FIG.2   HEAT-REMOVAL CUTTING CARBIDE INSERT

14.3 UNIVERSAL CUTTING TOOL WITH QUICK-EXCHANGE INSERT

There are many scientific works which has been devoted to the problems of designing of the universal insert holders for the machining processes (*Oberg, Erik.," Machinery's Handbook", 24th Edition, 1992, pp.697-699.-Industrial Press Inc., New York*) and (*Patent #360155.USSR, MPC B23, B29/04. Cutting Tool/Rozenblat A.I/USSR*).

The original attempts were made by author of this paper in question of designing firmly fixed carbide insert in holder of the cutting tool for turning and boring machining processes. This construction has the separate cutting block with carbide insert and holder which is joined by conical bolt permitting to change and fix the different cutting angles in the insert.

However, the presence of conical bolt at this construction does not guarantee the quick-exchange insert, particularly in the process of its breakage. The author tried to develop the new universal cutting tool which is shown in Figure 3 and Figure 4.

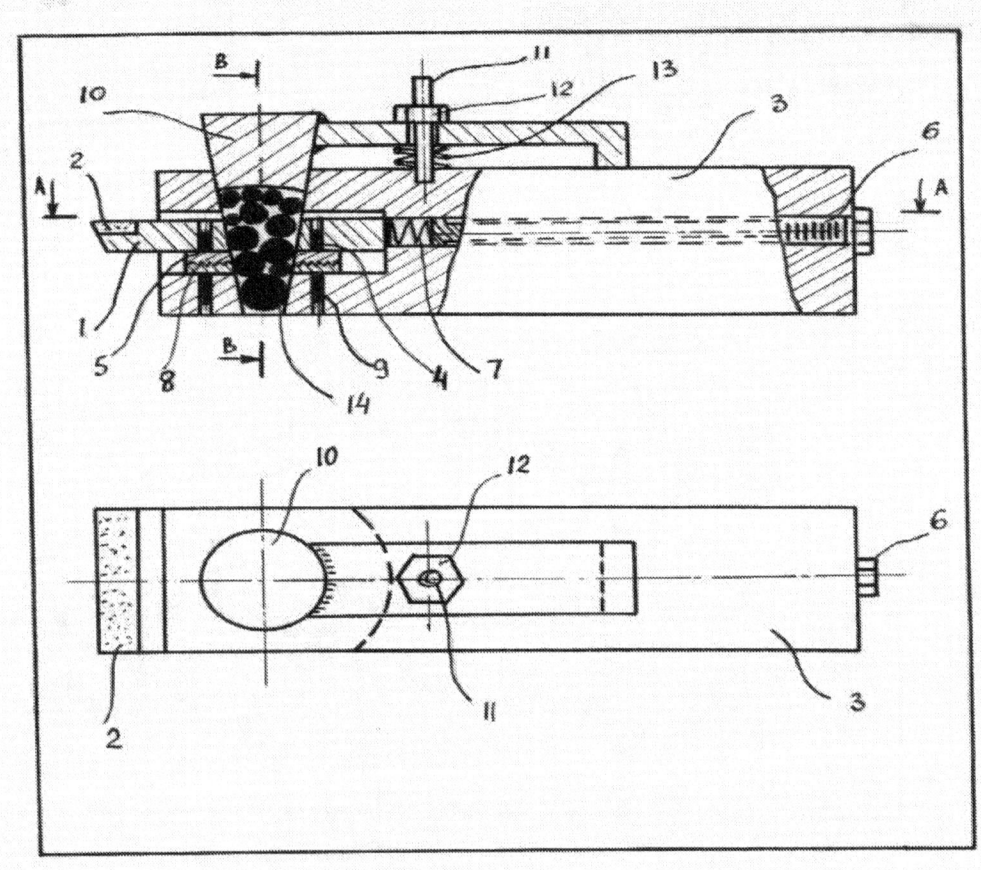

FIGURE 3   UNIVERSAL CUTTING TOOL WITH QUICK-EXCHANGE INSERT

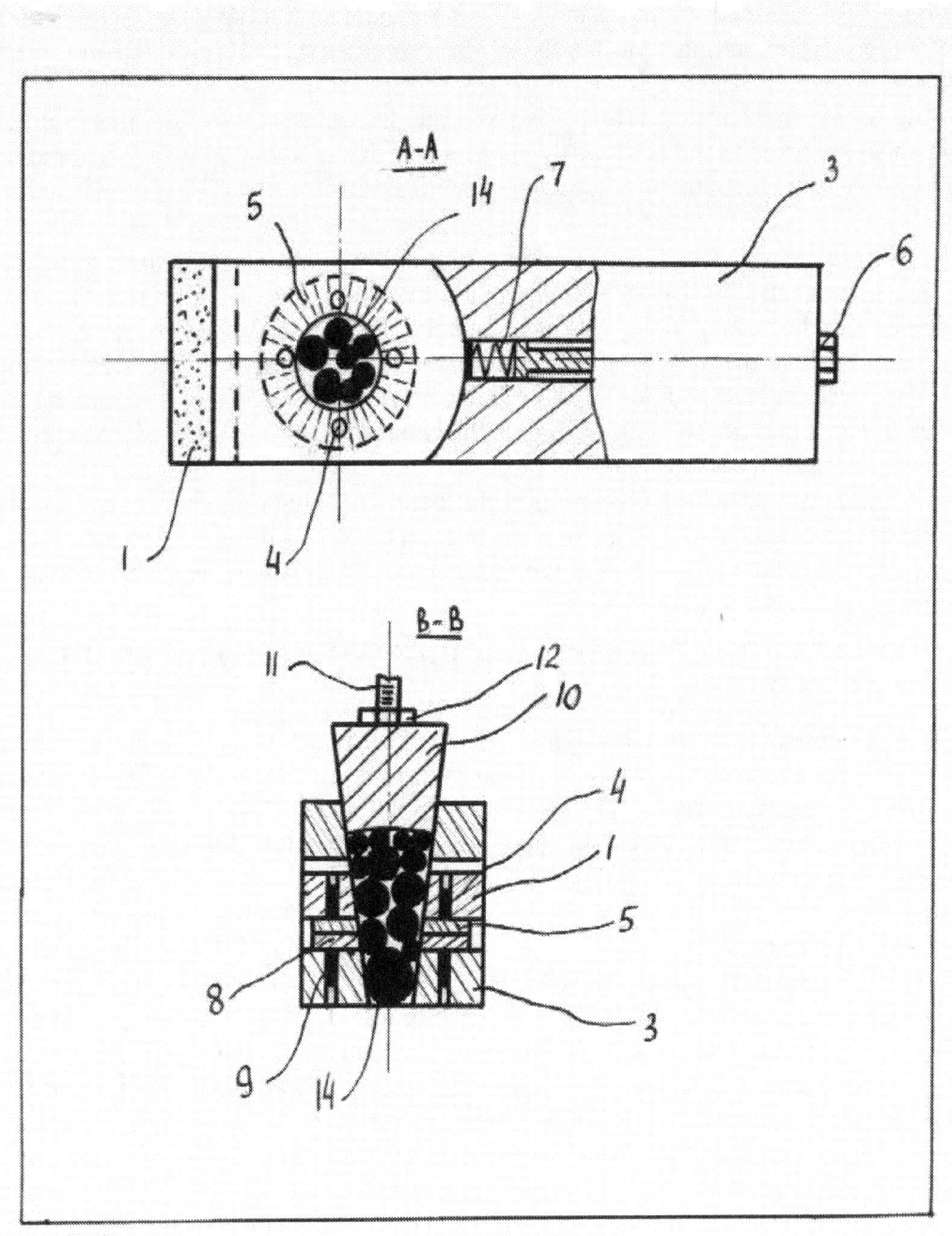

FIGURE 4   UNIVERSAL CUTTING TOOL WITH QUICK-EXCHANGE INSERT
(CROSS-SECTION A-A   AND B-B)

Universal cutting tool has the cutting block 1 with the insert 2 and holder 3. On the cutting block 1 is fixed with assistance of pins 4 the disc 5 having the metal teeth of a file. The holder of this cutting tool has the central hole in which is installed the screw 6 with spring 7.

And on the front of this holder 3 is fixed the disc 8 having also the metal teeth of a file which also are fixed to the body of this holder 3 with assistance of pins 9. On the upper part of this holder 3 is fixed the turning lever 10 with assistance of stud 11, nut 12 and spring 13.

- In process of adjustment of cutting tool and working process the cutting block with insert 2 fixes to the front part of holder 3 so that the teeth of a file of each disc 5 and 8 coincided and the cutting block 1 turns on the given angle in plane.

In conical hole the balls 14 of different diameters introduces in disorder and the turning lever 10 fixes in given position of the cutting block 1. After of these operations in body of holder 3 put the screw 6 with spring 7 for improving of stiffness all system of cutting process.

So, the presence at this construction the balls 14, which are introduced in disorder in conical hole the machinist permits quickly to exchange the cutting tool block 1 with insert 2.

And this considerably decrease the auxiliary time for adjustment and maintenance and the cutting process improves in whole.

14.4 LONGITUDINAL PROFILE MACHINNING BY ROTATED AND SHAPED CUTTING TOOL

The longitudinal profile machining (rotated turning) widely used in manufacturing processes (*Handbuch der Fertigungstechnik Herausgegeben von Prof. Dr.- Ing. Gunter Spur und Prof. Dr.-Ing. Theodor Stoferle, Band 3/1*).

However, this method does not guarantee the qualitative chip removal particularly for the work-piece with the irregular allowance from stainless and high-resistance steels. Special rotated turning tool suggested by the author with variable angles which removes the above-named disadvantages.

Figure 5 shows the schematic construction of rotated turning tool with variable angles. On the holder 1 of cutting tool is fixed the profile insert 2 which has the possibility to rotate around of it axis 3 by the special installation (not shown on drawing).

The profile insert 2 has on the front surface three different surface 4,5 and 6 which of them has angle of slope. For example, the front surface 4 is made with the angle of slope which is equal zero, and the front surface 5 has the negative angle of slope and the front surface 6 has the positive value for the angle of slope.

The profile of insert (or tool) 2 has the shank 7 for connection with driving for its rotation (not shown on drawing). Ba sic surface 8 of this insert 2 has contact with the surface of holder 1 of cutting tool. The cutting insert 2 is made in view of disc and has the back angle 9 on the periphery in total units with the front surfaces 4,5 and 6 the general cutting angles. And besides in some crossing places of the front surface 4 and 5 is made auxiliary back angle 10.

And in zone of crossing of the front surfaces 5 and 6 is made the auxiliary back angle 11. And in zone of crossing the front surfaces 6 and 4 is made the auxiliary back angle 12.

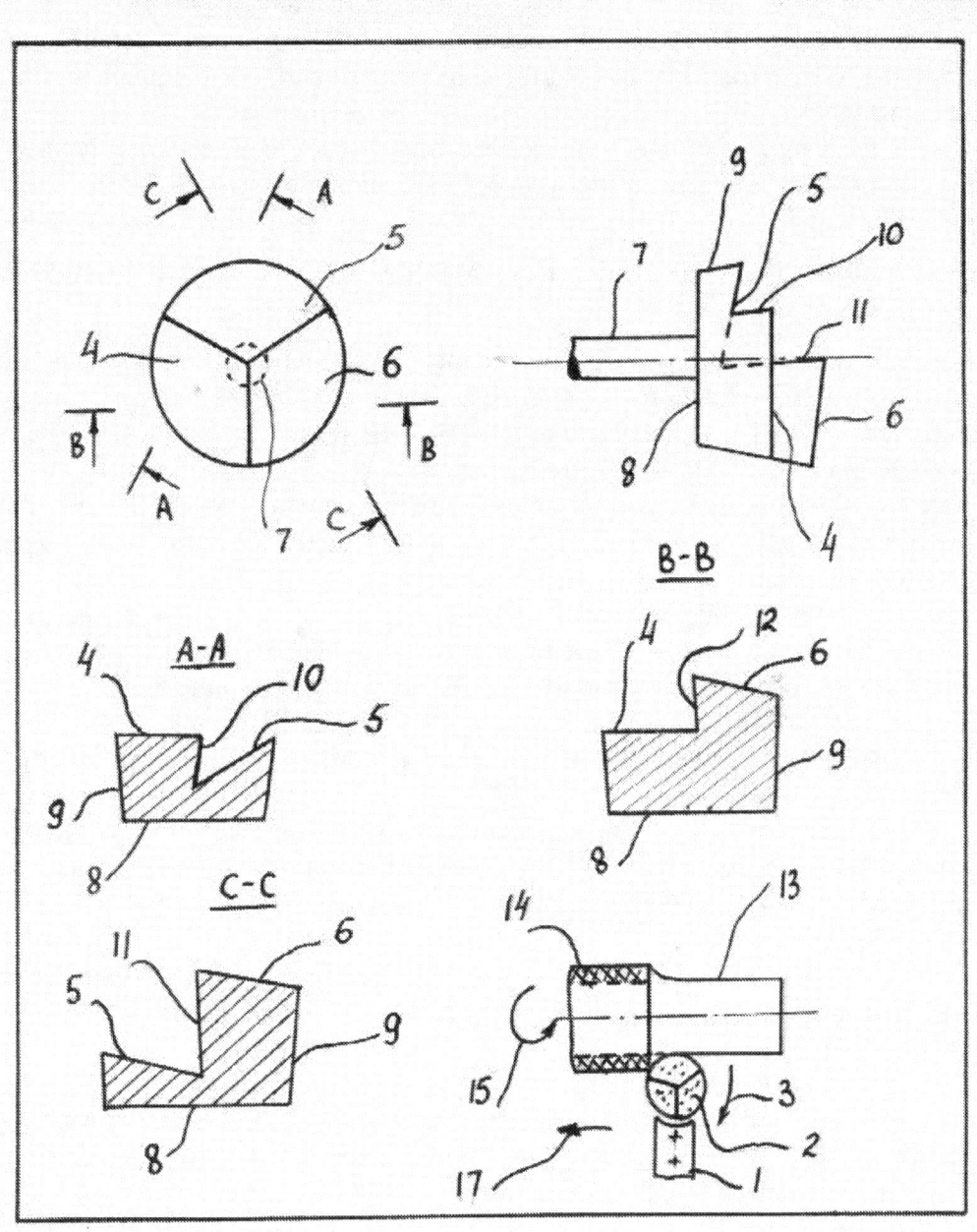

FIGURE 5 ROTATED TURNING TOOL WITH VARIABLE ANGLES

137

In process of cutting work-piece 13 with total allowance 14 the work-piece gives the rotation 15 and profile insert 2 also gives the rotation 16 and feed 17.

So, the work-piece 13 and profile insert 2 have both the different rotations and we see that from the work-piece 13 comes out the chip of small parts, i.e the process of break-chips goes in machining.

Owing to designing of the front surface of this profile insert 2 with three different surfaces improves the metal cutting process particularly of stainless steels and its break-chips.

14.5 PNEUMATIC TRANSPORTATION OF THE FERROMAGNETIC MATERIALS.

Pneumatic transportation of the ferromagnetic chip and dust play the important role in machining process (*Rozenblat A.I, Russian Inventor and Scientist brings the new technologies to USA ".- Proceedings, The 27 th Israel International Conference on Mechanical Engineers, Haifa, pp.476-482*).

Although there are many different most-known pneumatic transportation systems which are used in practice, but unfortunately they do not guarantee the durable for the pipe system and elbows.

The objective of this device is the increasing of the durable the pipe system advantageously of the elbows, as the most exposed to abrasive wear in process of removal the ferromagnetic chip and dust from the cutting zone of the machine tools.

The new pneumatic transportation system schematic is shown in Figure 6.

This system includes the dust-chip collector 1, horizontal pipeline 2 which is fixed by the element 3 and also elbow 4 and tank 5.

The peculiarity of this system that the elbow 4 has the possibility to rotate around of horizontal axis into account that on the periphery of this elbow 4 is fixed the cylindrical gear 6 which has the connection with leading gear 7 and driving 8 having the rotation from motor 9.

On the other end of this elbow 4 is fixed radial bearing 10 which is installed in body 11 of horizontal pipe 2. The tank 5 in upper part has the flange connection 11 and vibrators which are made in view of spring 12.

* In process of pneumatic transportation of chip 13 from cutting zone from the machine tool the last comes to the horizontal pipeline 2 and further to the elbow 4. And as the elbow has the rotation 14 around of his axis, the chip 13 under of actions of the centrifugal forces, comes to the tank 5 and further under of actions of the vacuum 15 (vacuum fan is installed outside of the building-not shown on the drawing), the chip 13 to the space of flange connection 11 and further comes to the vertical pipeline of pneumatic system.

In view of fact that the flange connection 11 is fixed with tank 5 by means of elements 16 and besides the tank 5 has the vibrator 12 we can do conclusion that the chip 13 constantly in outer surface of tank 5 comes in motion and has "hanged conditions".

In due of this fact the chip 13 removes from the tank by the forces of vacuum and further comes to the vertical pipe 11. In period of rotation of elbow 4 the last makes the complex motion and the chip 13 has the absolute velocity which is equal:

$$V_a = V_r + V_e$$

Where,

V_a = absolute velocity;

V_r = relative velocity from vacuum forces;

V_e = transfer velocity from rotation of elbow.

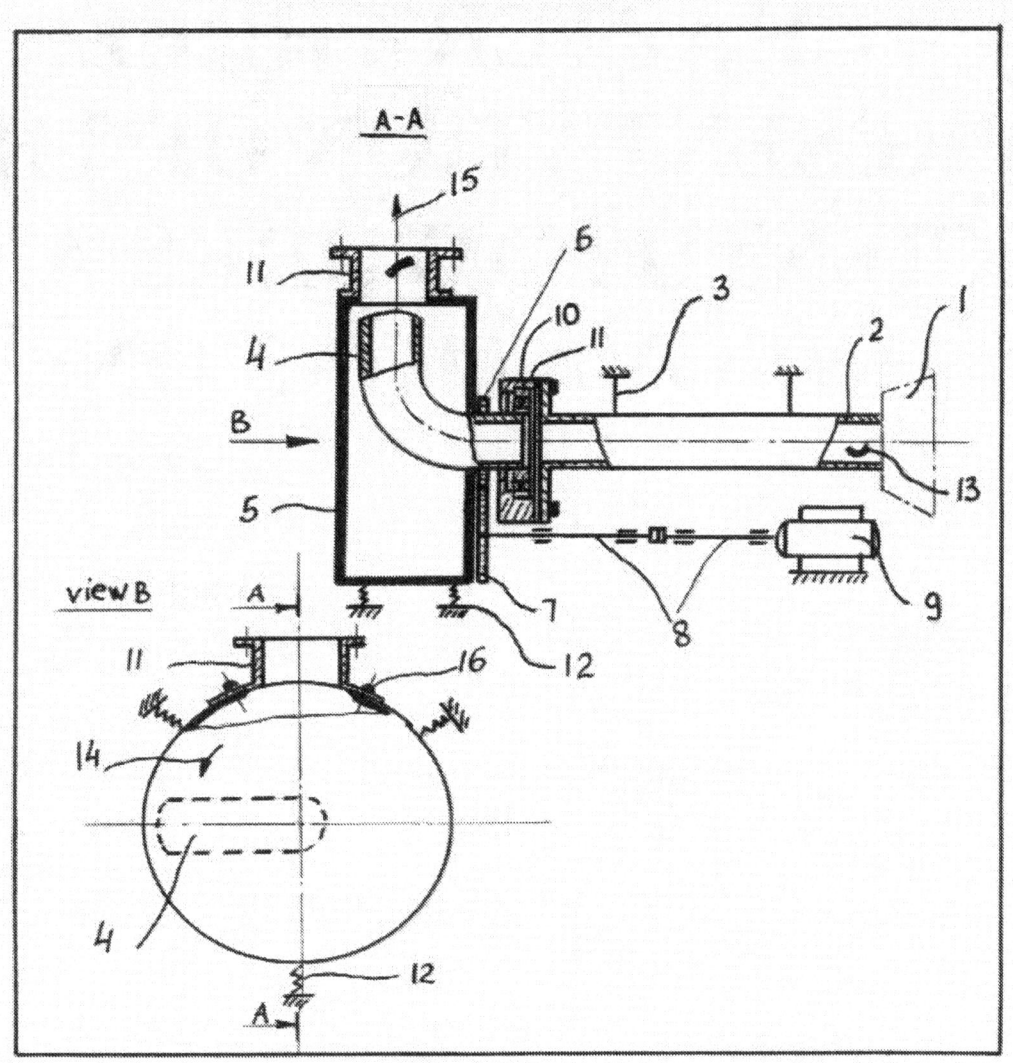

FIGURE 6   PNEUMATIC TRANSPORTATION OF FERROMAGNETIC MATERIAL

CHAPTER THREE
PRESENTATION OF THE SCIENTIFIC WORKS AND INNOVATIONS ON:

26 TH ISRAEL CONFERENCE ON MECHANICAL ENGINEERING

1. Isothermal method of transportation Liquefied Natural Gas (LNG) by maritime gas-carriers

ABSTRACT

The author explores the potential for export of liquefied natural gas (LNG) from production facilities in the former USSR to Western customers by sea, and proposes the method for isothermal LNG transportation by maritime gas-carries, in which low temperatures at standard pressures are employed to permit carrying the cargo.

In this paper is shown that this proposed isothermal LNG transportation method possesses advantages over pressured or combined methods of transporting. Designing of maritime gas-carriers, calculations regarding required refrigeration capacity, and the selection and evaluation of thermal insulation materials are treated.

Additionally, the author suggests a possible scheme for the direct cooling of LNG in the tanks of an isothermal gas-carrier, and considers different schemes for use of the secondary coolant gas. Attention is also given to matters of LNG handling safety.

INTRODUCTION

An analysis of the worldwide distribution of natural gas reserves yields the following shows that the USA is thought to posses $2.5 \cdot 10^{14}$ ft^3, Middle East $5.5 \cdot 10^{14}$ ft^3 and Sino Soviet block some $7.5 \cdot 10^{14}$ ft^3 [1]. Herewith the author [2] notes that annual domestic production of gas in the US is declining from the 1977.

Such conclusions are also supported by Drewry Shipping Consultants, Ltd.,(London) [3], indicating that U.S natural gas production will fall by about of 1.7 % per annual between 1990-1995, and by 1.5 % from 1995 to 2000. This information leads to the conclusion that a great potential for export of LNG from the former USSR to world markets, particularly the U.S.

TRANSPORTATION OF LNG BY GAS-CARRIERS:

BASIC DIRECTIONS

Based on the authors data [4], delivery of gas over distances greater than 1500 miles favors maritime methods, as the LNG gas-carrier offers greater flexibility of use, is less dependant on the nature and stability of freight traffic.

At present time, there have been developed three basic LNG gas-carriers:

a. Ships with reinforced tanks capable of withstanding the pressure of transporting gas, which has been liquefied by increasing pressure at normal temperatures (*adiabatic transport)*;
b. Vessel with thermally-insulated tanks for transportation of gas which has been liquefied by cooling it below it's boiling point (*isothermal transport)*;
c. Carriers with tanks both insulated and sufficiently strong to carry gas which has been both cooled and placed under pressure (*mixed transport)*.

It will become evident that transporting LNG at low temperatures and pressures close to one atmosphere has advantages over adiabatic transportation.

CONDITIONS SPECIFIC TO THE ISOTHERMAL METHOD OF LNG

In view of the fact that liquefying gas by cooling results in a bulk cargo of higher specific gravity than the LNG carried under adiabatic conditions, transportation of LNG by isothermal means yields greater given load-carrying capacity for ships than that obtained by carrying LNG under pressure. However, shipping cooled gases requires the installation of refrigeration equipment, high-power compressors, condensers, additional gas-lines and valves, and places a higherelectrical upon the ship than carrying pressurized LNG at ambient temperatures.

Considering in detail the physical-mechanical properties of metals at low temperatures for propane (-41.8° C at 1 Torr), for ammonia (- 33.5° C at 1 Torr) and butane (-0.5°C at 1 Torr) we see that for the manufacture of tanks for LNG (butane, propane and ammonia), it is possible to use stainless steel with 3.5 % nickel content. Requirements for the thermal insulation materials used to cover the tanks, considerations for the insulation used on both above—deck and below-desk tanks are relatively straightforward.

The material must not be flamed, steady against of moisture and the particular liquefied gas being carried, and have, of course, the lowest possible thermo-conduction.

It would appear that for these conditions, fiberglass in quite suitable, has a low cost and is relatively environmentally and demonstrates one acceptable version for covering tanks in on-shore situations.

There are some peculiarities unique to LNG transport at temperature near the boiling point of the gas that require changes to the construction of the body of the gas-carrier. As a rule, these vessels have double hull bottom and longitudinal bulkheads separating ship-borne water ballast compartments. The spaces between the tanks in during of transporting LNG, are flooded with inert gas (nitrogen or carbon dioxide) to reduce the likelihood of fire or explosion.

DESIGNING OF THE REFRIGERATION PLANT OF LNG TANKER

An important matter to be considered in isothermal LNG transport this is the quantity of heat input to the cargo of refrigerated LNG. It is important to know the amount of such input that is likely to occur so that the refrigeration plant may be sized properly.

In the matter of heat transfer into the LNG, one may readily use the classical thermodynamics and empirically derived engineering data. The over-all coefficient of heat transfer U for the walls of the cargo tank for the gas-carrier can be determined, in the main, as for a laminated flat wall by the formula:

$$U = 1 / (1/f_1 + L/ K_f + 1/ f_2) \quad (1) \, [\, 5\,]$$

where,

K_f = thermal conductivity for insulation;

L = thickness of insulation for walls of cargo tank;

f_1 = film coefficient (or heat-transfer coefficient)from inert gas filing interstitial space to the outside surface of the cargo tank wall;

f_2 = film coefficient (or heat-transfer coefficient) from the inner wall surface of the cargo tank to the liquid propane.

To determine f_1 use the NUSSELT number [Nu]:

$[\, Nu \,] = (f_1 \cdot h) / K_f \,(2)$

where,

h = height of wall plate;

and also the GRASHOF number $[\, Gr \,]$:

$[\, Gr \,] = g \cdot h^3 \cdot \beta \,(\Delta T) / \upsilon^2 \,(3)$

where,

ΔT = the temperature difference between wall and inert gas;

υ = the kinematics viscosity is equal $\upsilon = \mu / \rho$

β = the coefficient volume-metric expansion;

g = the gravitational acceleration;

ρ = the mass density of inert gas;

μ = the viscosity of inert gas.

Substituting in Eq.(3), we get

$[\, Gr \,] = h^3 \cdot \rho^2 \cdot g \cdot \beta \cdot (\Delta T) / \mu^2 \,(3\,a)$

And besides we must use the PRANDTL number $[\, Pr \,]$:

$[\, Pr \,] = \upsilon / a \,(4)$

where,

a = the coefficient thermal diffusivity is equal

$a = K_f / C_p \cdot \gamma$ or

$[\, Pr \,] = C_p \cdot \gamma \cdot \upsilon / K_f \,(4\,a)$

C_p = the specific heat capacity of inert gas;

γ = the specific weight s equal $\gamma = \rho \cdot g$

And besides we must calculate the REYNOLDS number $[\, Re \,]$:

$[\, Re \,] = \rho \cdot h \cdot V / \mu \,(5)\,[\, 6 \,]$

where,

V = velocity of inert gas.

Considering the film coefficient f_1 as the complex value we can write

$[\, Nu \,] = C \cdot (Gr \cdot Pr)^n \,(6)$

where

C and n experimental values C =0.54, n= 0.25.

Equation (6) may than be written as

$(f_1 \cdot h) / K_f = C \cdot [(h^3 \cdot \rho^2 \cdot g \cdot \beta \cdot (\Delta T)) / \mu^2 \cdot (C_p \cdot \mu \cdot g) / K_f \,]^n \,(6a)$

and

$f_1 = [\, (C \cdot K_f) \cdot h \,] \cdot [\, h^3 \cdot \rho^2 \cdot g^2 \cdot \beta \cdot C_p \cdot \mu^{-1} \cdot K_f \cdot (\Delta T) \,]^n \,(7)$

where,

$f_1 = C^* \cdot [\, \rho^2 \cdot \beta \cdot C_p \cdot \mu^{-1} \cdot (\Delta T) \,]^{0.25}$ and $C^* = 0.54 \cdot h^{-0.25} \cdot K_f^{0.75} \cdot g^{0.50} \,(7\,a)$

Equation (7) shows a functional relationship between the film coefficient f_1 and the variables n the form

$$f_1 = \varphi\,[\ h,\ K_f,\ \rho, \beta,\ C_p,\ \mu,\ \Delta\ T)\ ]\ (8)$$

For determining of coefficient f_2 take into account that liquid natural gas boils in the tank at the atmosphere conditions, we have the following equation:

$$f_2 = 6.9 \cdot 10^{-3} \cdot [\,(\gamma^{11} \cdot r)\,/\,(\gamma^{1} - \gamma^{11})\,]^{1.0} \cdot [\,(\gamma^{1}\,/\,\sigma s)\,]^{0.33} \cdot [\,(\lambda^{0.75} \cdot q^{0.7})/(\mu s^{0.45} \cdot Cp^{\,0.12} \cdot Ts)^{0.3}\,]$$

where, (9)

$\gamma^{\cdot\cdot}$ =specific weight of steam natural gas;

$\gamma^{\cdot}$ = specific weight of liquid natural gas;

T_s = absolute temperature saturation;

r = heat of evaporation;

σ_s = superficial tension of liquid natural gas;

λ = coefficient of heat conduction liquid natural gas;

q = specific heat load in area of film and bubble boiling;

C^{1}_p = specific heat of liquid natural gas;

μ_s = coefficient viscosity of liquid natural gas.

The determination minimum thickness of insulation walls tank calculate owing to the impossible of falling out dew to the formula:

$$(L\,/\lambda) = (1/\,f_{1)} \cdot [(t_v - t_{p)}\,/\,(t_v - t_{r)}]\ -\ (1\,/\,f_{1)} - (1\,/\,f_{2)}\ (10)$$

where,

t_v = temperature of external air;

t_p = temperature of keeping liquid natural gas;

t_r = temperature of falling out dew.

At the determining of coefficient K_c heat transfer construction insulation deck usually uses the following formula:

$$K_c = (K_f \cdot \Phi)/\,S\ (11)$$

where,

S = width between longitudinal deck beam;

Φ = form-factor taking into consideration thermal influence of set.

METHODS FOR COOLING AND SECONDARY LIQUEFACTION OF LNG

Even f it were possible to construct and LNG tank with 100 % perfect insulation refrigeration would still be required to prevent evaporation of the liquid gas. As a minimum, refrigeration would be required to overcome heat input to such a degree as to keep the gas temperature below the boiling point.

Such a minimal refrigeration plant would save weight and space which could then be given over to cargo. However, due to kinetic heating of the sloshing LNG in a tank aboard pitching ship, economy could only be realized by using such a small cooling apparatus for short voyages.

Evaporative loss can be avoided by liquefying the gas directly in the tank. Figure 1 shows some possible schemes of such cooling.

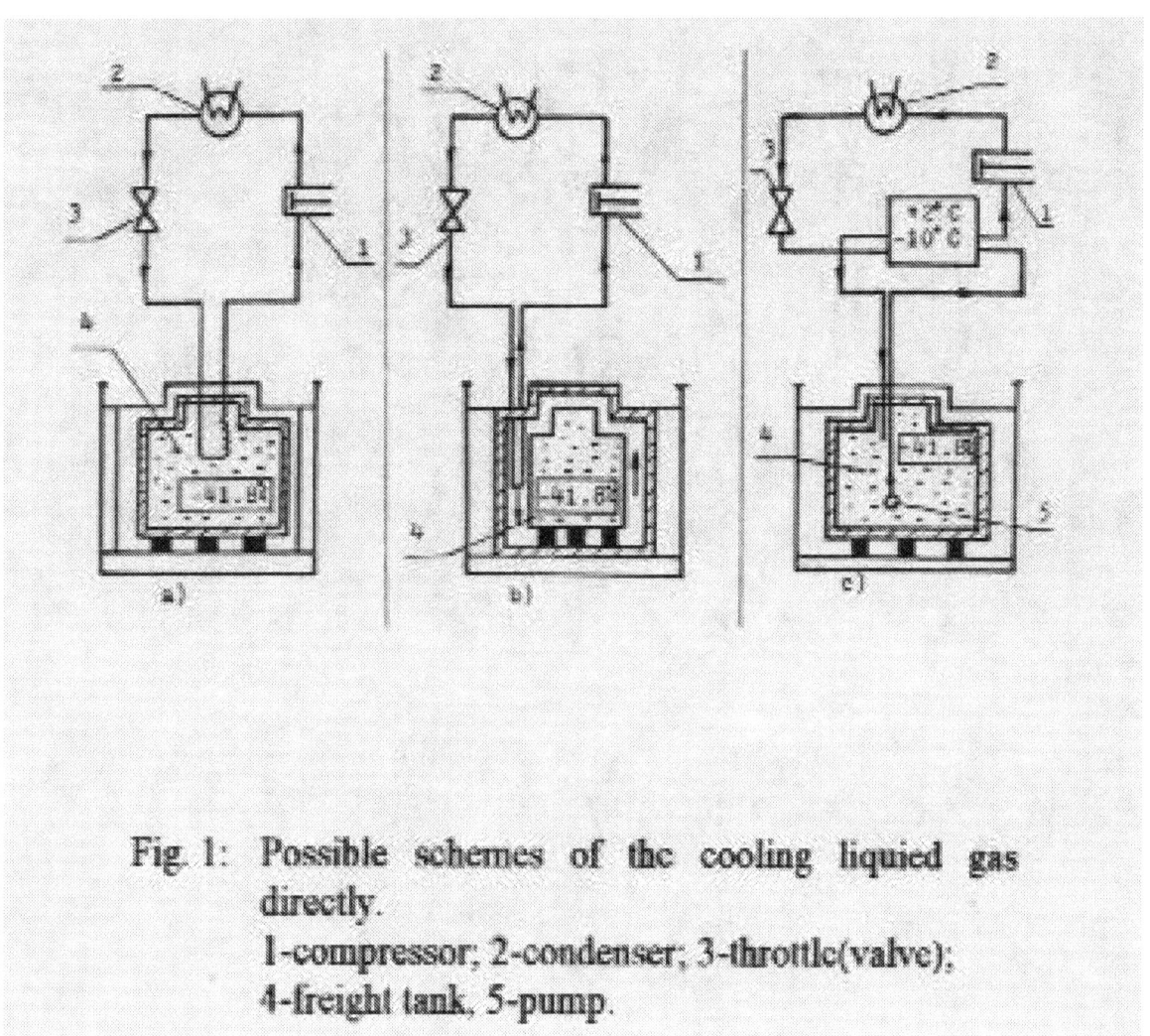

Fig. 1: Possible schemes of the cooling liquied gas directly.
1-compressor; 2-condenser; 3-throttle(valve);
4-freight tank, 5-pump.

Method (a) provides for cooling the cargo gas by a coiled pope installed within the tank.

Method (b) employs the coolant coils applied to the exterior of the tank.

Method (c) pumps the LNG to an external heat exchanger where the unwanted heat is removed.

The previously discussed methods from reducing evaporative freight loss are impractical due to their excessive energy consumption. More economical would be a secondary liquefaction of evaporated gas and it's subsequent return to the storage tank.

Figure 2 shows possible schemes for the secondary liquefaction of is evaporated gas.

In part (a), the LNG which has evaporated is the first compressed (1), then flows through a condenser (2). The liquid LNG, under pressure and at a temperature above the boiling point, then passes through throttle-valve (3) which performs the role of reducing the pressure to a point where the expanding gas cools sufficiently to condense.

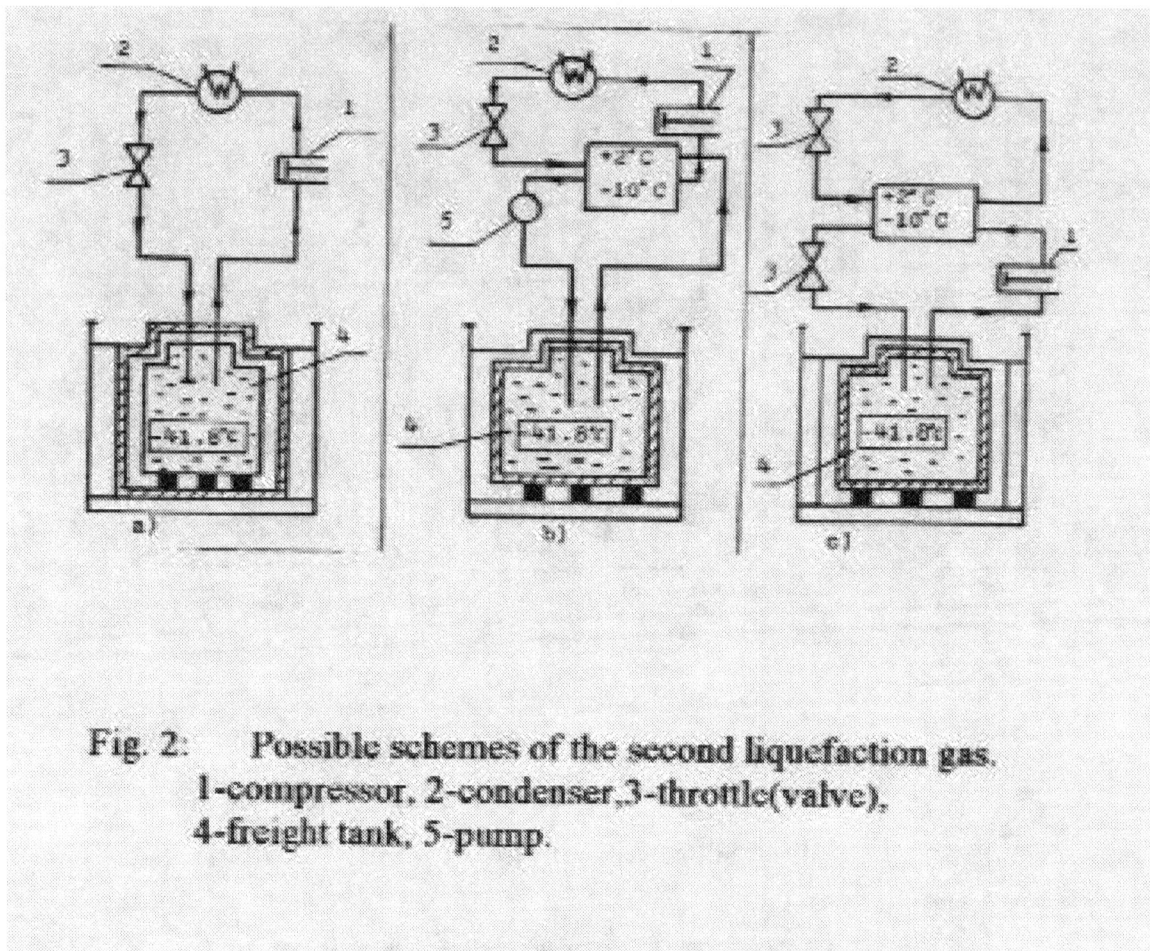

Fig. 2:   Possible schemes of the second liquefaction gas.
1-compressor, 2-condenser, 3-throttle(valve),
4-freight tank, 5-pump.

The liquid gas, stripped of heat acquired in the tank, the returned to the storage tank.

In scheme (b) evaporated gas is passed through a refrigeration plant which carries off the heat of evaporation. The gas is thus liquid and pumped back into the cargo tank.

Finally, at **scheme (c)** we find a scheme similar to scheme (b), but the evaporated gas liquids before coming to the condenser and is then under pressure; after of expansion via the throttle valve (3) and corresponding heat loss, the gas, now liquid and at 1 Torr is returned to the storage tank. The heat carried by the gas is drawn of through the external refrigeration apparatus.

*The methods just described each have their own advantages and disadvantages. The storage temperature for each of the gases of interest is:

Propane (- 41.8°C);

Ammonia (- 33.4° C);

Butane (- 0.50°C)

From the schemes described above, the **method (c)** with a two-gradient refrigeration plant where ratio of working pressures is $P_k / P_0 = 11.05/1.05 > 8.00$ which permits flexible enough temperature control to cool any of the gases of interest to below it is boiling point.

LOADING AND UNLOADING EQUIPMENT FOR MARITIME LNG GAS-CARRIERS

As compared to carriers designed to transport LNG under pressure, ships carrying this cargo under isothermal conditions require the following equipment:

I.   In each tank set, a submersible pump, with drive mounted safely outside of the tank. This pump delivers the LNG to the main centrifugal pump on deck, which transfers the cargo to shore side storage tanks. Before loading, the tanks and piping must be purged of air by flushing with inert gas as, shown in Figure 3.

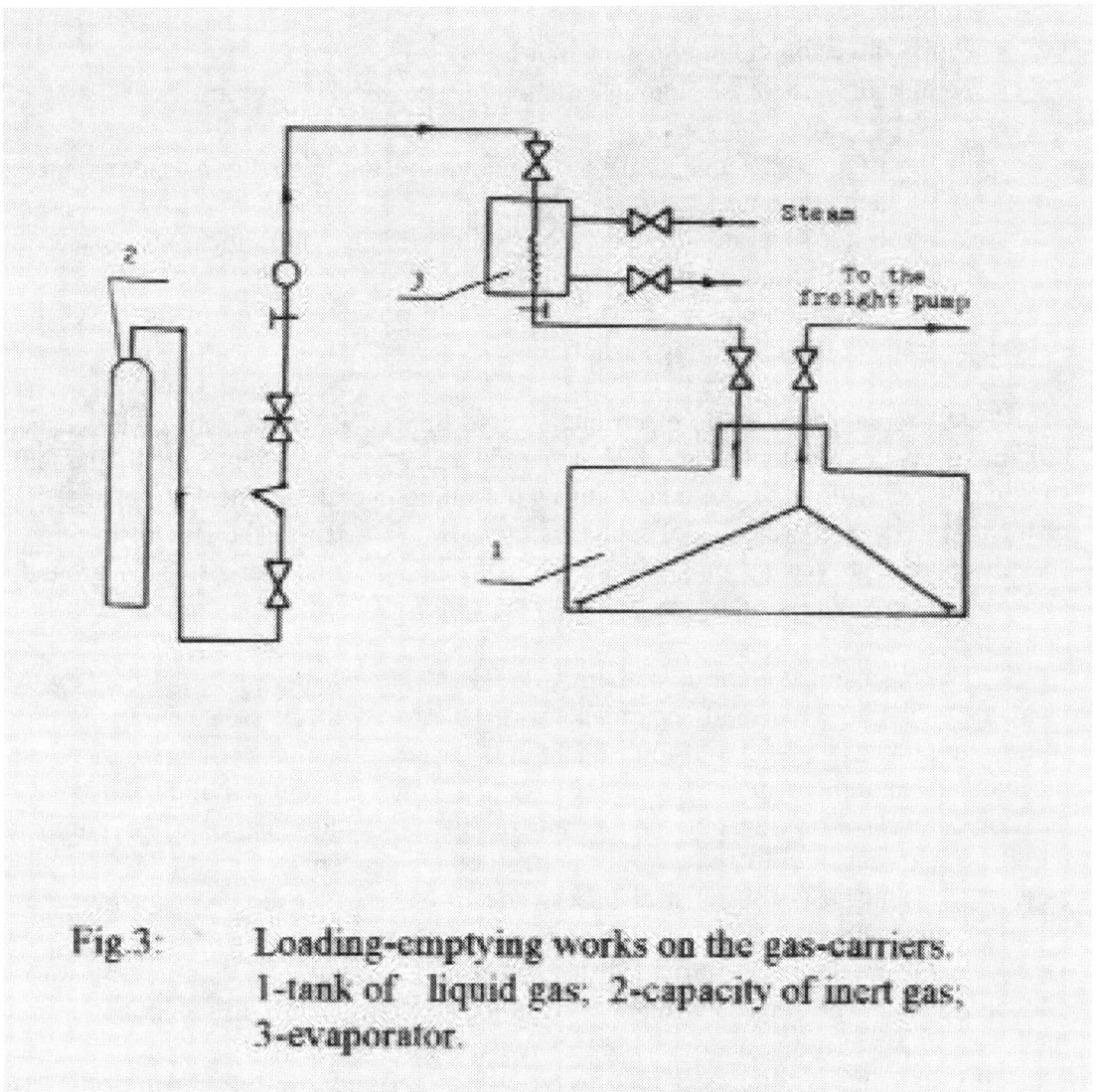

Fig.3:   Loading-emptying works on the gas-carriers.
1-tank of liquid gas; 2-capacity of inert gas;
3-evaporator.

This purging procedure must be followed for the sake of safety any time the tank is opened for inspected, during loading and unloading.

II.   After of purging, the tanks need to be cooled to their working temperature (i.e., below the boiling point of the gas to be carried). For this objective, an apparatus which

Anatoly Rozenblat

allows the inside for the tanks to be sprayed with liquid gas may be used. Such a system consists of:

    a) Liquid—level indicator of type ECVC—SHK;
    b) Two safety valve (working and controlling) to guard against excessive tank pressure caused by:
        * heating of the tank by any fire which might break out on the ship;

- An increase in the volume of the due to an increase in temperature beyond the normal volume of 85 % full;
- Pressure-relief valves to protect against excessive evaporation of LNG which might occur at the breaking of the pipelines, knocking-out of fittings, or misalignment of gasket, etc.
- Globe valves on the liquid and vapor lines;
- To measure loss of LNG an installation of types UST-40 is retried consisting of:
-

    1. Liquid—volume register of round—piston type, to measure total and delta volumes;
    2. Filter-separator to clear LNG of debris and impurities, and to separate LNG liquid from vapor;
    3. A differential valve which is used to prevent boiling of the LNG in the volume-measuring system.

III. Consideration of static electricity is also important. Movement of LNG in the pipeline generates a potential of 3 KV or so, and a charge is also generated by the falling stream of LNG within the cargo tank during the filling operation. The static charge could result in sparking, the danger of which is obvious. To reduce the risk of ignition by spark, suitable provisions must be made to ground off the charges. This grounding procedure is especially important prior to connection of loading piping and tank filling or emptying.

IV. The mechanical plant in isothermal LNG carriers usually placed in the bow well away from living areas, and separated from the rest of the vessel by of appropriate watertight bulkheads. Also, the atmosphere in the areas of pumps, compressors and other mechanical equipment must be changed frequently to prevent a buildup of gases to explosive levels. For this purpose, a rate of 40 air-changes per hour in pump and compressor rooms and 20 changes per hour in other mechanical facilities is considered safe. In general, provision must also be made to ensure that air does not move from pump rooms to other areas.

V. Suitable means be provided to detect the presence of gas in the various compartments of the ship. Gas analysis equipment exists which can detect and monitor the levels of gas in living quarters, pump rooms, etc.

The safe limits areas follows: **Propane (2.30 %); Butane (1.50 %); Ammonia (16.5 %).** Maintenance of gases at these levels or below will result in save conditions.

SUMMARY:

1. We have outlined three methods of maritime LNG transport: high pressure, isothermal and combination of two.

2. Isothermal method of LNG transportation has several inherent advantages over transport under pressure. Isothermal method require tanks of lower weight, can result in greater volume—use efficiency and lower transportation costs.

3. It seems that if a two—stage refrigeration system, following the pressure relationship $P_k/P_0 = 11.05/1.05 > 8.00$, yields the best results in keeping the LNG under the proper conditions for isothermal transport.

4. Isothermal transport can be safety accomplished, providing the appropriate measures for inert-gas purging, ventilation and equipment grounding are taken.

REFERENCES:

[1] R.L. Loffners, Energy Handbook. (New York: Van Nostrand Reinhold Company), p. 46 (1978).

[2] D.J. Cuff, W.J.Young, The United states energy atlas. (New York: The Free Press A. Division of Machillan Publishing Co., Inc.),46, (1980).

[3] Prospects for LNG and LNG shipping in the 1990s. (London: Drewry Shipping Consultants Ltd., England), 43, (1990).

[4] S. M. Logachev, The gas-carriers for transportation of Liquefied Natural gases (Moscow, Publishing House. Maritime Transport, (1965).

[5] J.F. Lee, F.W. Sears, Thermodynamics. (Massachusetts: Addison-Wesley Publishing Company, Inc.),337, (1963).

[6] Mark's Standard Handbook for Mechanical Engineers. (McGRAW-Hill Book Company,9th Ed.),3-49, (1987).

2. ANALYSIS OF CHIP FORMATION IN CUTTING STAINLESS STEELS WITH CIRCULAR SEGMENTAL SAWS

ABSTRACT

In the suggested article, the author examines one important problem of circular sawing such as chip formation in the process of cutting, mainly big rolled products from stainless steel, by cold circular segmental saws.

Analysis of data has shown that geometrical parameters of chip have accidental characteristics and depend on the conditions of machining, form of work-piece and submitted to the laws of mathematical statistics.

In the article, a big attention is given to the analysis of distribution of the geometrical parameters of chip (external and internal diameters, number and steps between wraps of chip, and width and thickness of chip).

Herewith marked that the shape of frequency polygon of geometrical parameters of chip are distributed as abnormal distribution—*Platykurtic.*

Geometrical analysis of stainless chip will allow in perspective to use the different arrangements for its breaking and withdrawing from the cutting area by systems of pneumatic transportation.

INTRODUCTION

In industry widely use stainless and heat-resistance steels. However, process cutting of blanks advantageously from stainless and heat-resistance steels for rolled products and forging of a big diameter by circular segmental saws arise definite technological difficulties which are joined with low productivity of machining from the insignificant resistance of tool [1].

The utmost importance on the rising of the resistant tool has learned very well the question of chip-formation methods, its breaking and withdrawing from the cutting area. But these questions conformed to the circular segmental saws have learned insufficiently and particularly in evaluation of types and forms generated chip and also construction of its sizes [2].

Evaluation of characteristics chip makes access to use different arrangements for its breaking and withdrawing from the cutting area particularly by systems of pneumatic transportation [3].

EXPERIMENTAL PART

For analysis in industrial condition, was selected the austenitic chrome-nickel-titanium steels Cr18N9T (*analogous with stainless steel 302 AISI* [4]) in view of round bars with diameter D=180 mm (7.09 in.). Cutting of material has been fulfilled by well-known methods on the circular sawing machines by circular segmental saw with geometric parameters:

- external diameter of saw $D_e = 710$ mm (27.95 in.);
- width of saw W = 6.5 mm (0.26 in.);
- quantity of teeth of saw Z = 96 in the following regimes of machining cutting:
- *speed V =13.26 m/min (43.50 fpm);*
- *horizontal feed on the teeth of saw f_t = 0.03 mm/teeth (0.001 in/teeth);*
- *minute of feed f_m = 150 mm/min (5.91 ipm).*

Period of resistant of circular segmental saw has been estimated by the methods of the author [5]. Shape of teeth circular saw and sharpening of it has been made in accordance with recommendations of the author [6] on the special semi-automatic grinder with the following control of quality of sharpening, as shown in FIGURE 1.

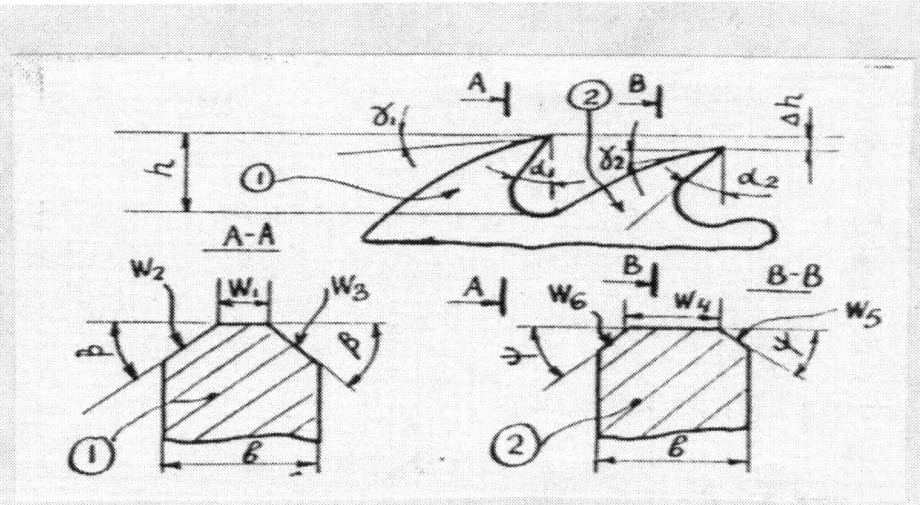

Fig. 1: Shape of circular segmental saw and sharpening of its
teeth:
1-rough-slotting teeth; 2-finish-scraping teeth.

[clearance $\gamma_1 = \gamma_2 = 8°$;

  rake $\alpha_1 = \alpha_2 = 12°$;

                    $\triangle h = 3.0mm$ (0.012in);
                    $w_1 = 3.0mm$ (0.118in);
                    $w_2 = w_3 = 2.5mm$ (0.098in);

  chamfer $(\beta = \Psi = 45°)$;
                    b =6.5mm (0.26in);
                    $w_4 = 5.0mm$ (0.197in);
                    $w_5 = w_6 = 1.0mm$ (0.039in);
                    h =10.0mm (0.394in).]

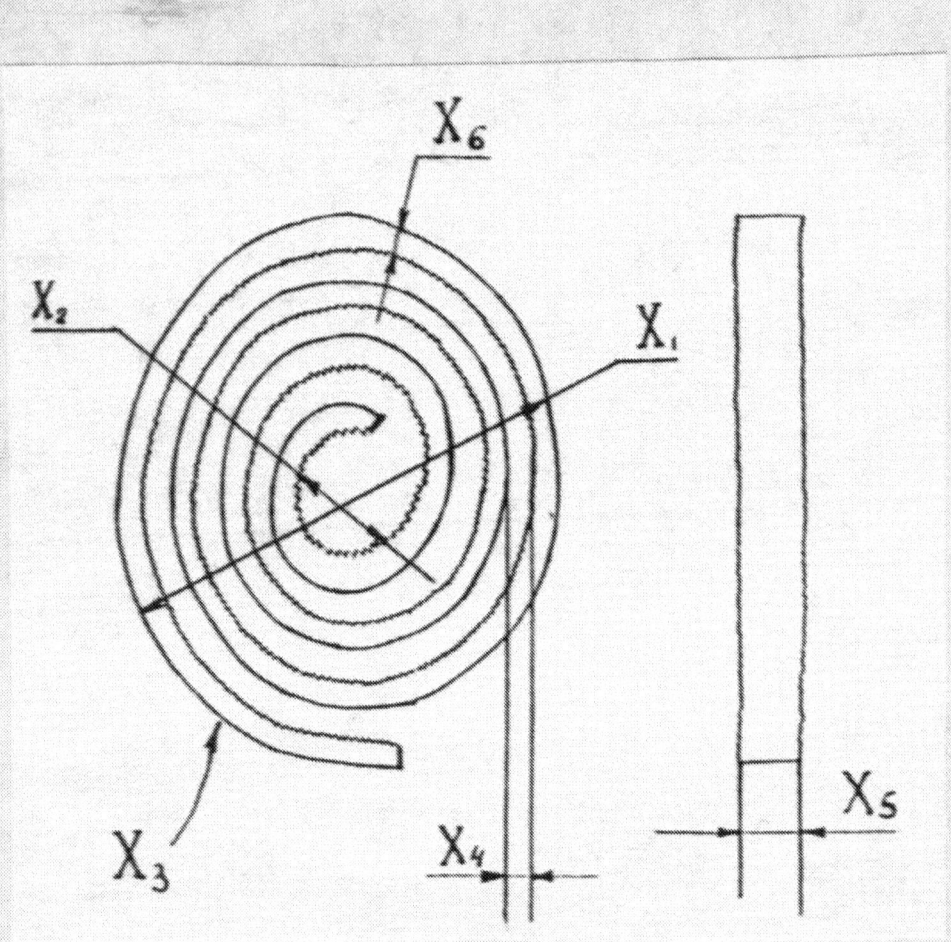

Fig. 2: Constructive sizes of chip are formed in
period of cutting stainless steel by circular
segmental saw.

(X₁-external diameter; X₂-internal diameter;
X₃-number of wraps of chip; X₄-step between
wraps of chip; X₅-width of chip; X₆-thickness
of chip).

DISCUSSION AND CONCLUSIONS

The experimental data of constructive sizes of chip were obtained from the total number of observation N=42. Graphical method of data presentation is shown in FIGURE 3.

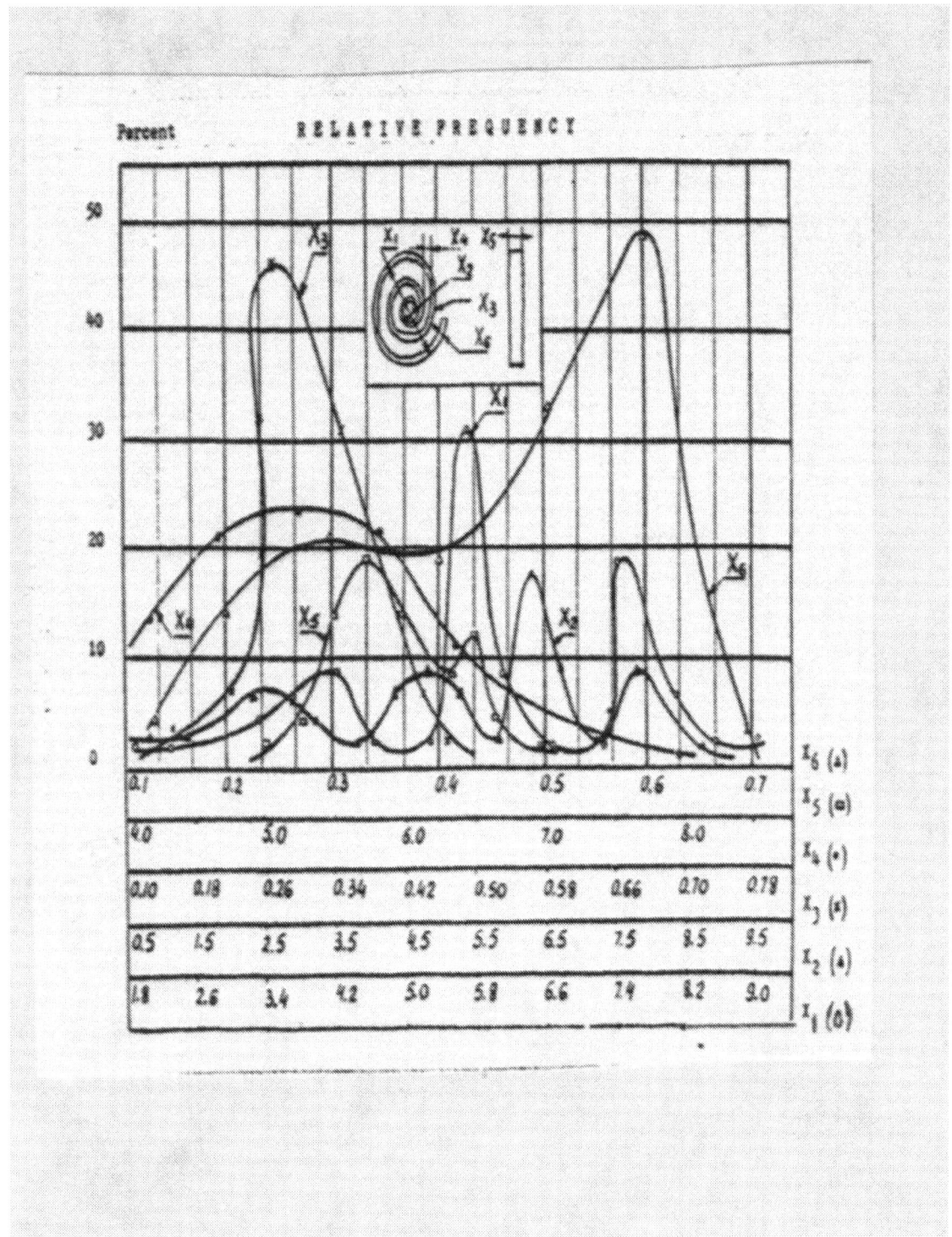

Figure 3 Frequency polygon distribution for the geometrical values of chip at the cutting stainless steels with circular segmental saws

The most important characteristics of distribution are summarized in Table 1 for the constructive sizes of chip.

Table 1 Summary characteristics of distribution for stainless chip

Symbolism	Parameter of distribution	Size of chip					
		X_1	X_2	X_3	X_4	X_5	X_6
Mw	Mode	10.05	5.70	2.75	0.28	6.10	0.52
Mdnw	Median	10.11	5.60	2.90	0.28	6.22	0.36
Xw	Mean	10.18	5.50	2.96	0.36	5.84	0.40
S.Dw	Standard deviation	0.40	1.86			0.70	0.19
CVw	Coefficient of variation	3.66	12.34			5.60	0.21
g1	Skewness	0.24	0.00	0.07	0.42	0.00	0.00
g2	Kurtosis	-1.81	-1.57	-1.39	-0.92	-1.28	-1.57

As we in FIGURE 3 presented the abstract results of statistical analysis in view of relative frequencies of polygon distribution in evaluation of constructive sizes for stainless chip.

Analysis combined distribution of the values $_1$ show that the shape of frequency polygon of data is distributed as abnormal distribution- PLATYKURTIC.

Combined distribution of the values X_2 displayed in Figure 3 show that the shape of frequency polygon of data is distributed also as abnormal distribution- PLATYKURTIC, but having a lower percent of relative frequency.

The curve of the values X_3 characterizes asymmetrical distribution with positive skew (Skewness g_1 =0.07) and slightly PLATYKURTIC (Kurtosis g_2 = - 1.39). Analysis of the values of X_4 shows that curve has the shape of distribution which characterizes asymmetrical positive skew distribution with skewness (g_1=0.42) and slight PLATYKURTIC distribution (Kurtosis g_2= - 0.92).

Analysis combined distribution of the values X_5 from FIGURE 3 show that the shape of frequency polygon is distributed as abnormal—PLATYKURTIC. The finally the polygon of distribution the values X_6 displayed in FIGURE 3 show that the values seek to disperse from center and this shape of distribution characterizes as PLATYKURTIC with value of Kurtosis g_2 = - 1.57.

SUMMARY

1. Chip formation in the process of cutting mainly big rolled products from stainless steel by cold circular segmental saws has a great value with point view of learning its shape and geometrical parameters of chip (external and internal diameters, numbers and steps between wraps of chip, width and thickness of chip).
2. Making a careful close study of geometrical parameters of chip in cutting process with cold circular segmental saws we can in perspective use the different arrangements for its breaking and withdrawing from cutting area by systems of pneumatic transportation.
3. Geometrical parameters of chip have accidental characteristics and depend on the condition of machining, form of work-piece and submitted to the laws of mathematical statistics.
4. Analysis description of distribution of geometrical values of stainless chip show that frequency polygon of data is distributed as abnormal distribution—PLATYKURTIC.

REFERENCES

[1] A.I.Rozenblat, Criteria of evaluation of resistance circular saws in during cutting of billets. (Kiev), 40-42, (J. Technology and Organization of Production),10, (1974).

[2] Metal-cutting: Today's Techniques for Engineers and Shop Personnel.(American machinist, Ed.),130, (1977).

[3] Author's Certificate # 1131634, USSR, MKI3 B 23 Q 11/02 (Moscow),35, (Bulletin" Discoveries. Inventions"),48, 1984.

[4] Machinery's handbook.(Industrial Press, Inc., N.Y,24 Ed.),373, (1982).

[5] A.I.Rozenblat. Analysis of resistance of segmental circular saws.(Moscow),16-18, (The collections of papers" Machinery Technology",3,(1975).

[6] Amstead, B., Ostwald, P. Metal cutting process. (John Wiley & Sons, Ed.), 596, (1977).

[7] Havilcek, L. Crain, R., Practical statistics for the physical sciences. (American Chemical Society, Ed), (1988).

[8] Schmid, C., Statistical graphics. (Wiley-Inter-science Publication), 1983.

Anatoly Rozenblat

27TH ISRAEL CONFERENCE ON MECHANICAL ENGINEERING

1. Rozenblat's new cutting tools for manufacturing processes

ABSTRACT

This paper concerns the mechanical engineering advantage of machine processing and contains some new innovations in the area of cutting tools, progressive dies and methods which will be useful in the manufacturing process today and in the future.

The main attention of the author is devoted to questions of methodology in designing new progressive tools and equipment, and also of its technical-economical indexes.

Also this paper shows the general directions in the process of designing the special cutting and auxiliary tools.

Major attention was given by the author to the problem of increasing tool life and also decreasing material and power capacity, particularly on the cut-off machines.

In detail were investigated the questions of designing cutting tools for the break-chips processes, particularly in the cutting of stainless and heat-resistance steels.

INTRODUCTION AND BACKGROUND

Mechanical engineering at present reached considerable progress in the design of new cutting tools and technological processes [1] and [2].

However, there are generally unsettled questions such as:

1. Metal rate consumption and workshop costs are very considerable in the manufacture of universal cutting and boring tools for machining operations;
2. Manufacturing cost of work-piece in some technological operations of cut-off machines and die-forging presses is very high;
3. Tool life of cold circular segmental saws for the cut-off operations is very low, particularly for bars with big diameters and made from stainless or heat-resistance steels;
4. The technological possibilities of progressive dies are not enough for increasing the productivity and decreasing of manufacturing cost;
5. The org-economical characteristics and provision do not take into account the designing of cutting and auxiliary tools for machinists who have some physical disabilities such as blindness, deafness, etc.
6. For work-pieces of heavy engineering industry (crankshaft, rotor, etc.) with large machining allowances, there is the problem of designing a special cutting tool;
7. There is the problem of breaking the long-strip chips, particularly in the machining operations of stainless and heat-resistance materials.

Some attempts at removing the above-named disadvantages were made by many authors [3] and [4], but all are not enough for modern manufacturing processes.

THE GENERAL DIRECTIONS IN THE DESIGNING OF PROGRESSIVE AND EFFECTIVE CUTTING TOOLS

1. *The decrease of metal rate consumption and manufacturing cost in production of cutting tool.*

   At present in manufacturing industries, machining operations widely use the right and left hand turning tools and some of its different modifications [5].

However, the author suggests the new cutting tool [6], as shown in Figure 1 which allows the machinist to change the side cutting edge angles and to decrease the metal rate consumption.

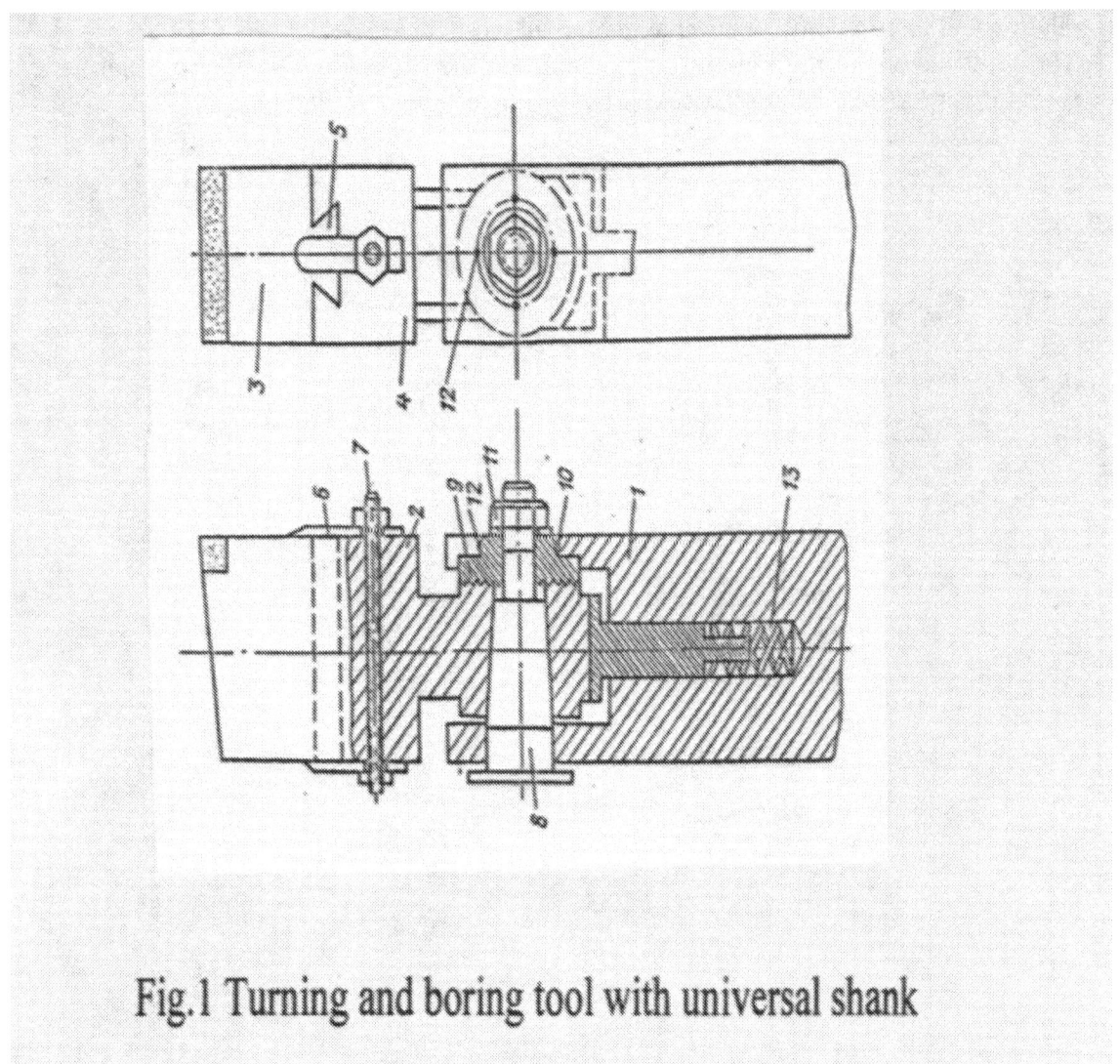

Fig.1 Turning and boring tool with universal shank

This tool is made with a separate head and shank. The head is fixed in shank of said lathe tool on its conical axis so that it is tightened in the conical hole of the head and fixed in given position by means of a screw.

2. *The decrease of power consumption in manufacturing processes.*

This factor is more important for the cut-off processes that use the cold circular segmental saws.

Application of the new Combined installation of A.I. Rozenblat [7], as shown in Figure 2, converts the kinetic energy of the free-falling cutting part from the bar into useful energy (work).

The combined installation is comprised of a milling-cutting machine for cutting the bar (placed on the upper deck), and stamping press for piercing or punching, (fixed on the lower deck).

After a heavy bar is cut-off, it falls, due to the force of gravity, down into the special installation and the stamping press, making useful work.

The arrangement of milling-cutting machine on the upper deck of the installation, and the stamping press, on the bottom deck, provides the efficient work.

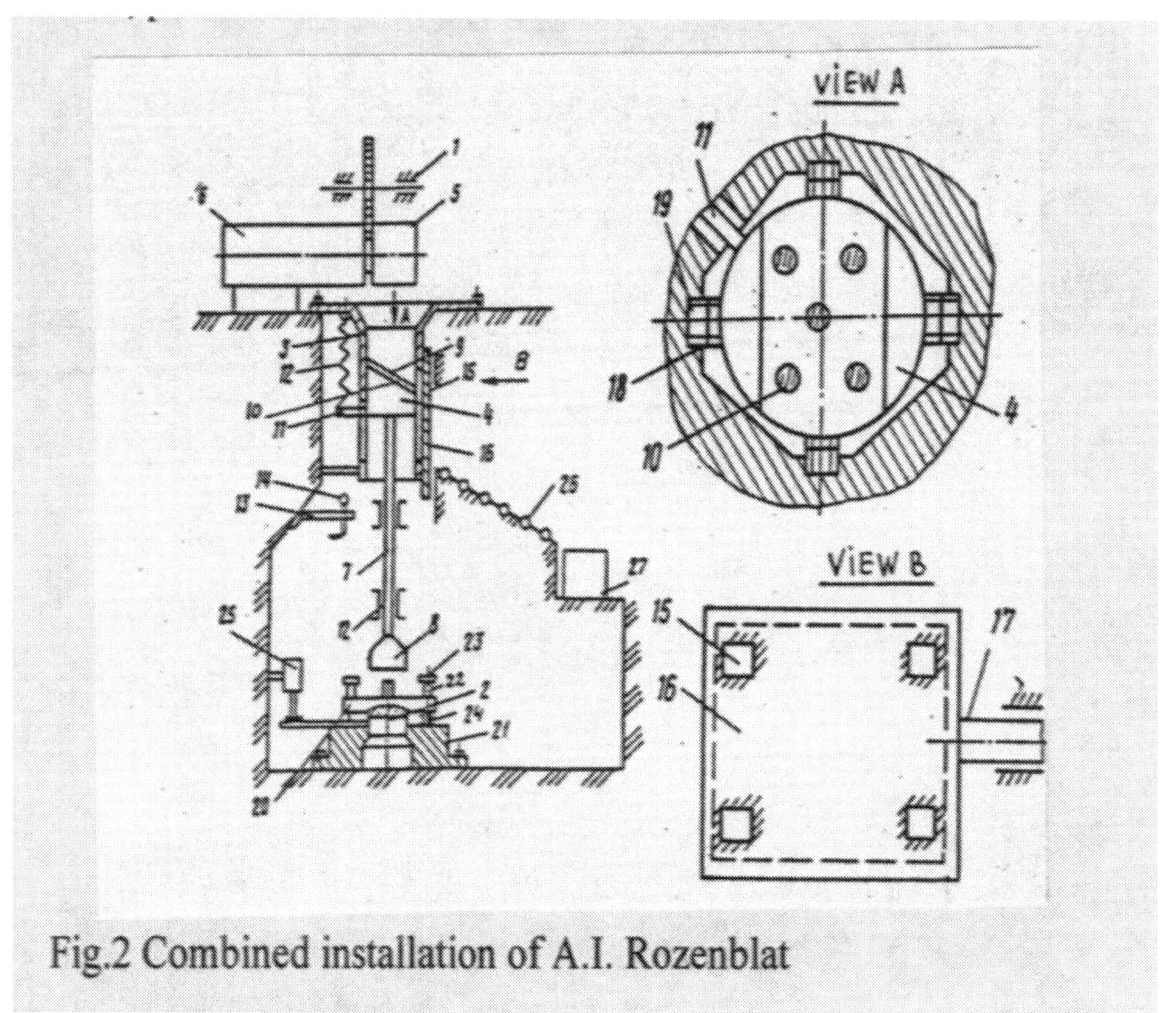

Fig.2 Combined installation of A.I. Rozenblat

3. *The increase of tool life for the some cutting tools.*

The cold circular segmental saw and parting-off cutting tool are suggested by the author, as examples of tools, that can have their tool life increased by a new device [8], as sown in Figure 3.

The main objective of the tool cleaning device is to automatically remove chips or fragments (e.g of metal from a cutting or milling operation) from the zone of the teeth, particularly in the operation of milling or cut-off machines.

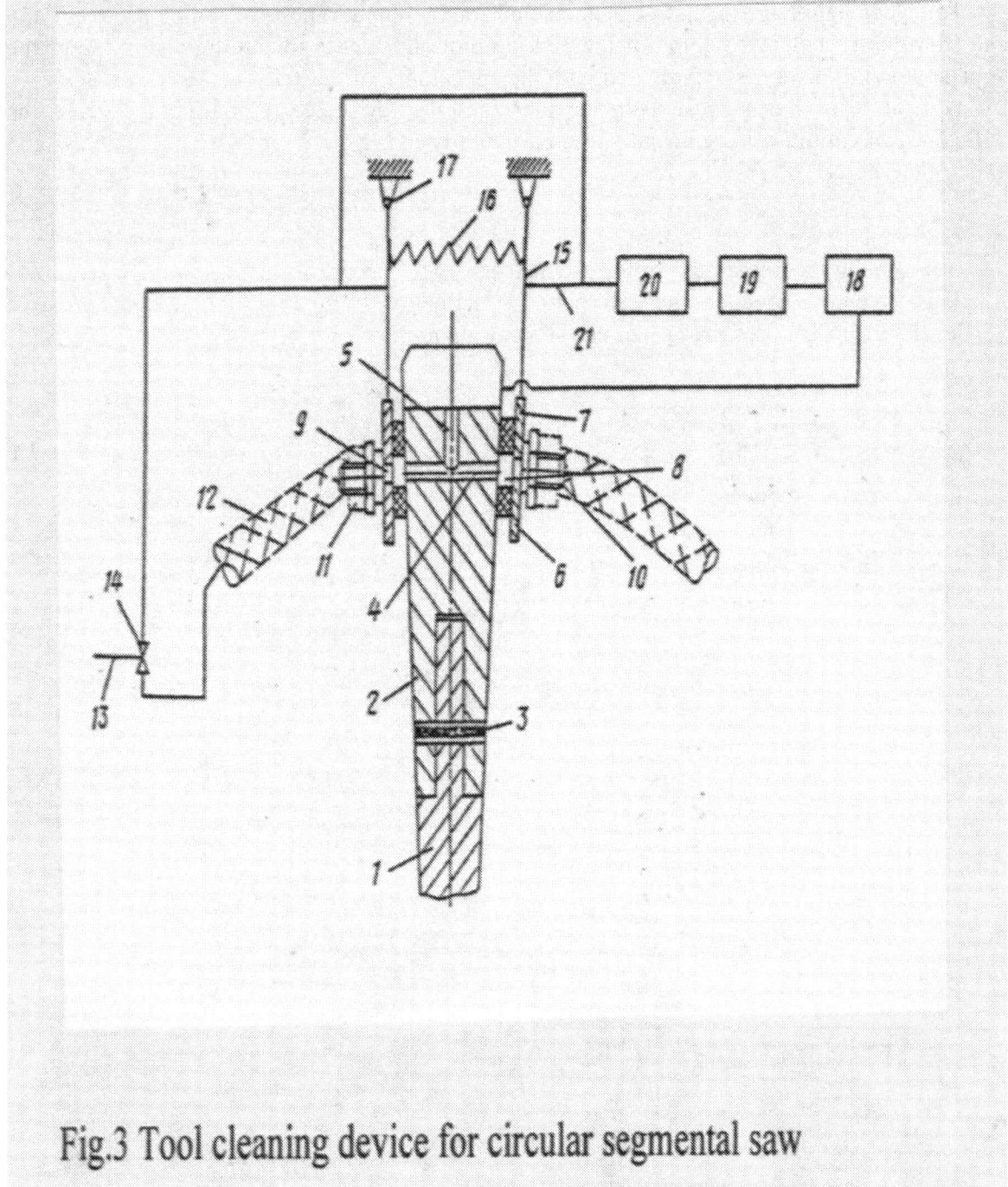

Fig.3 Tool cleaning device for circular segmental saw

This device uses two spring-actuated bars in opposing reciprocal motion, placed over the faces of a tool. The tool itself is given specific "cleaning grooves" along its transverse and longitudinal axis.

The cleaning device then sweeps chips and fragments into these "cleaning grooves" and out from the tool. The actuation of the pick up motion is set to match the revolution of the tool, so that it is always synchronized with the appearance of the "cleaning grooves".

Utilization of the suggested device increases the resistance of cutting tools and the efficiency of cutting-off segmental saws, particularly in the cutting of viscous, stainless and other materials.

- The other assembling cutting tool of A.I.Rozenblat **[9],** as shown in Figure 4, considerably decreases the fretting which normally occurs in metal cutting operations. This invention produces an automatic reduction of fretting by providing alternative levels of support/pressure at the rear of a cutting tool.

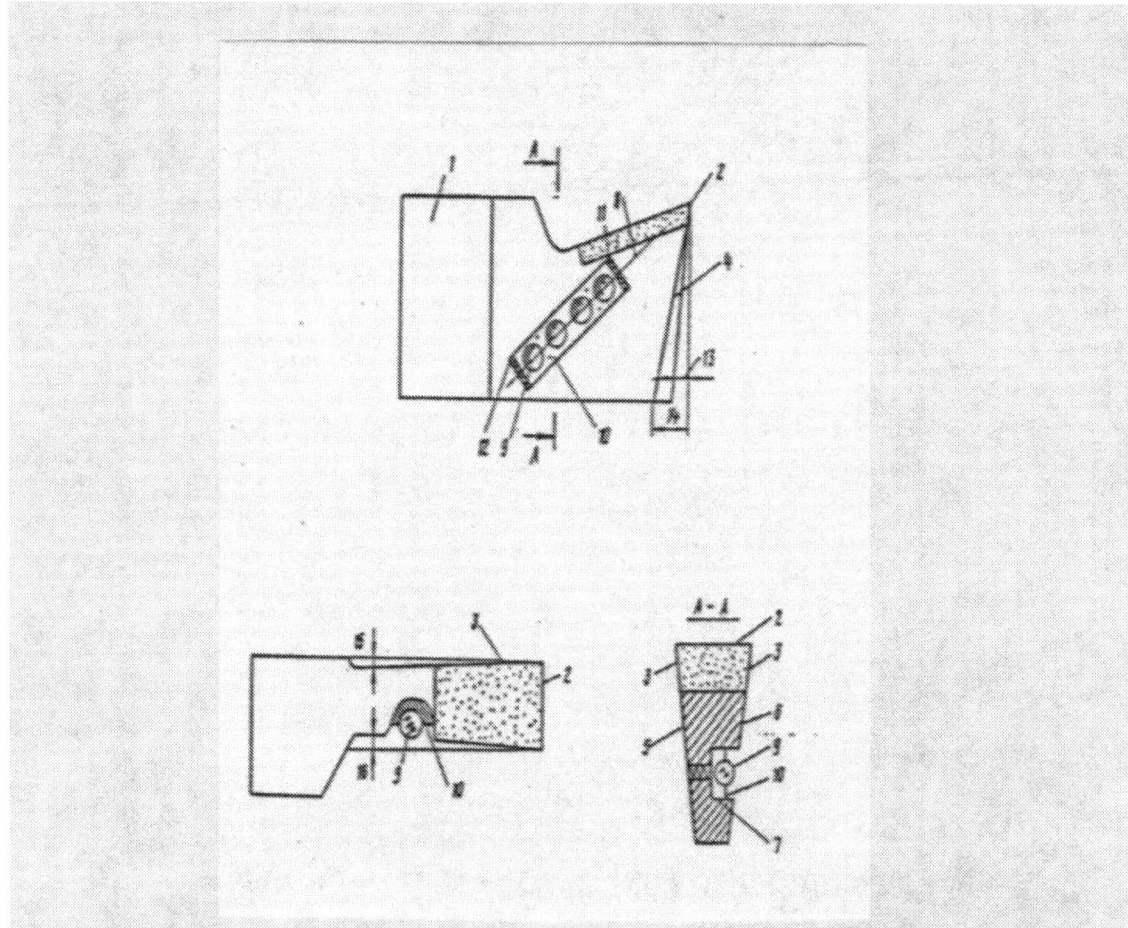

Fig.4 Reduced-fretting assembling of cutting tool of A.I. Rozenblat

This is done by means of a tool holder with rolling-contact bearings. As the cutting edge is fed longitudinally and wears slightly, it is pressed into an auxiliary rear surface and comes into contact with one of the rolling-contact bearings.

This provides the additional support that reduces the fretting as the cutting surface wears. So, by changing the coefficient of friction (iron-iron f=0.44) for the coefficient of rolling (also iron-iron with f_1=0.0005 m) on the suggested device, the tool life increases considerably.

4. The expansion of technological possibilities of progressive dies in condition of the flexible production.

Conformably to flexible production, a more recommended device is the multi-operational combined die of A.I.Rozenblat **[10],** as shown in Figure 5.

This combined die is comprised of at least two pairs of punches and dies for divisional operations. They are installed into the axis of press, so that one pair of punches is movable while the other die is unmovable.

The die on the other side and the punch connected between them is a total unit and is installed between the movable and unmovable parts of die, having elastic elements situated between the corresponding pair of punch-die.

The original construction of this device has movable and unmovable parts so, that some dies move horizontally and some punches move in the vertical direction.

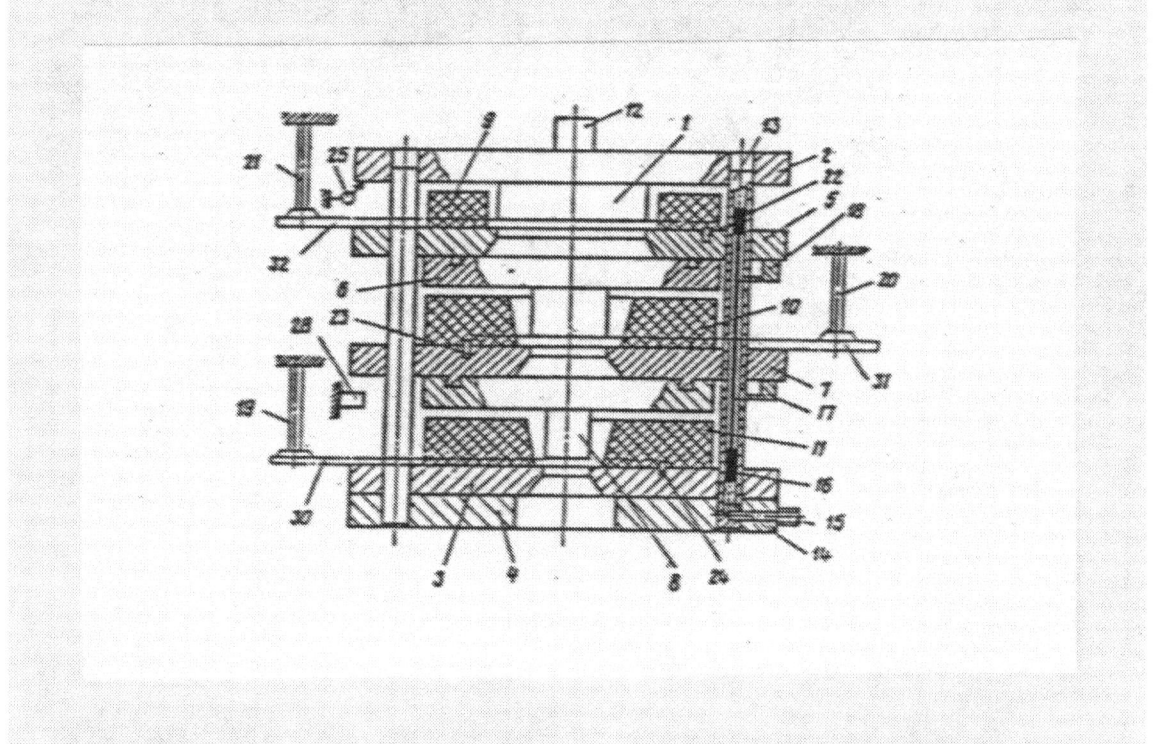

Fig.5 Multioperational combined die of A.I. Rozenblat

- The next device is a construction of a multi-operational die with cone, as shown in Figure 6.

This multi-operational die has the movable punch which is made in the form of a truncated pyramid. On the perimeter of truncated pyramid are joined the four movable horizontal punches: two of them displaced according to the first punches.

And besides this movable punch have additionally the fifth punch which is fixed tightly. So, for one movement (feed) of the press, the multi-operational die makes five or more different stamping operations.

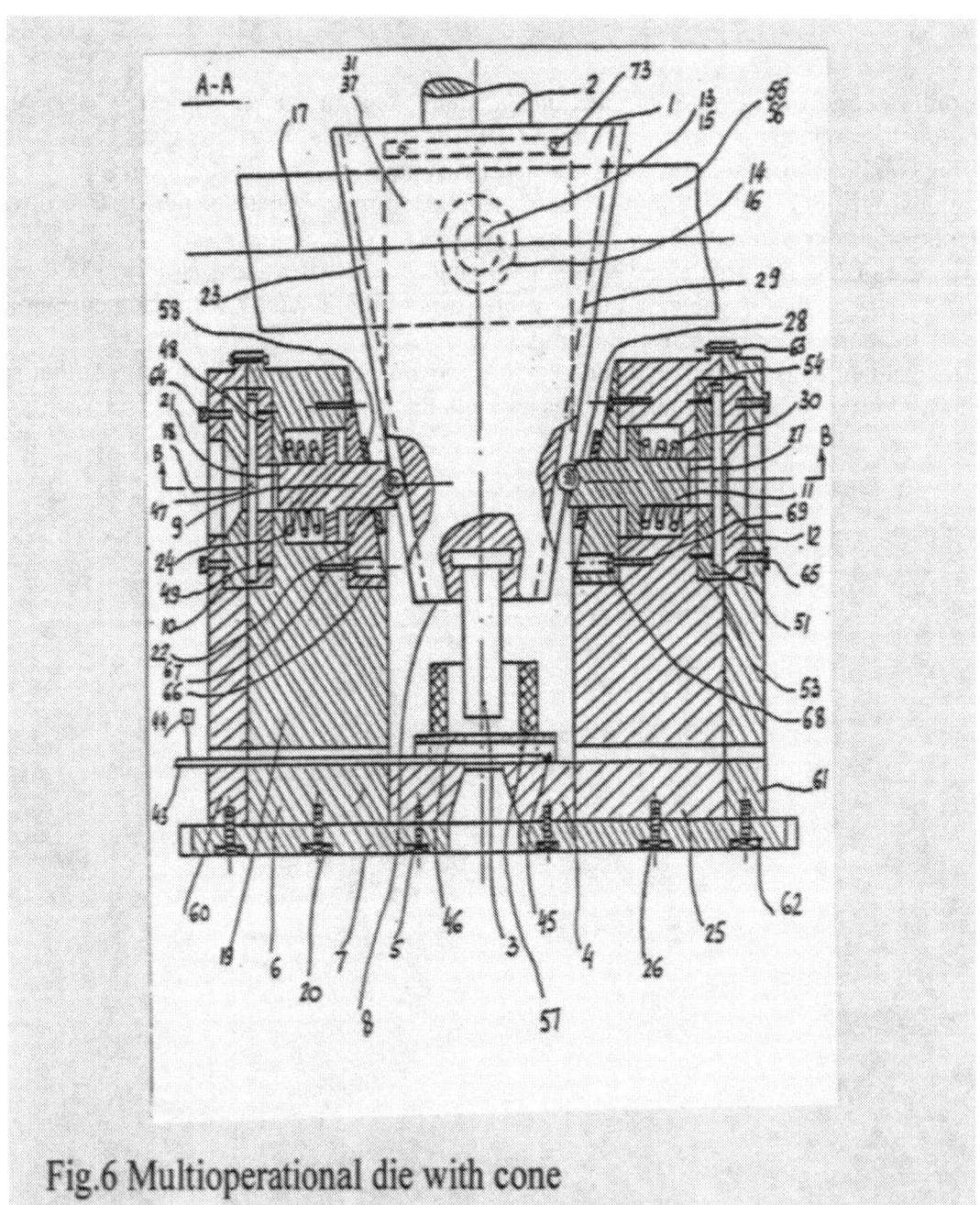

Fig.6 Multioperational die with cone

- To the varies of multi-operational dies relate Rozenblat's other combined die, as shown in Figure 7.

It includes some pair punches and dies which produce the working process at the straight movement of press (the punch comes down) and other pair punches and dies which produce the working process at the back movement of this press (punch comes up).

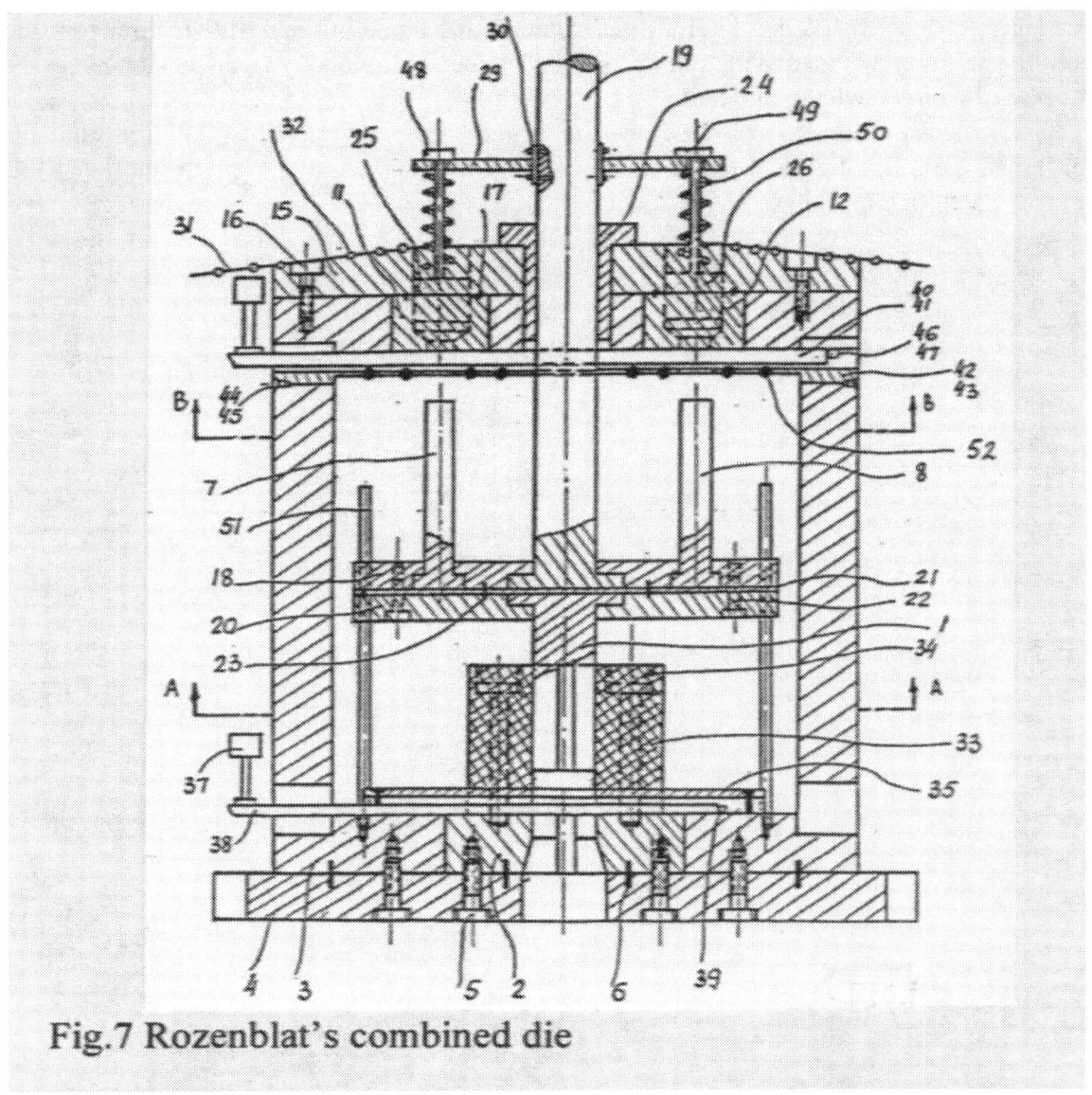

Fig.7 Rozenblat's combined die

5. The improvement of erg-economical characteristics in the designing of new tools

Improving the erg-economical characteristics in the design of new tools for machine processing is the main objective of this device.

The first step of this concern is the problem of designing safety tools for the blind or deaf-mute machine tool operators.

An example of this type of designing is a wrench of A.I. Rozenblat [11] for the jaw chuck, as shown in Figure 8.

The wrench for the jaw chuck supplies unified elements of electrical circuits with sources of power, signal installation, and interrupter of circuits, mechanism action on interrupter.

The handle and rod are made from two parts. The power is installed in both parts of the handle. The interrupter and signal installation are fixed on the part of the rod, while the mechanism action or interrupter is on the other part of the rod.

Thus the wrench has the possibility of rotation and made in view of the spiral pair and belt transmission, with a driving pulley used for connecting the body to the jaw chuck in the process of clamp-unclamp of detail.

Safety guarantees are provided by the sound and light signal which prevents the machinist from leaving the wrench inside of the jaw chuck. Also all system closes at this time and stops the machine.

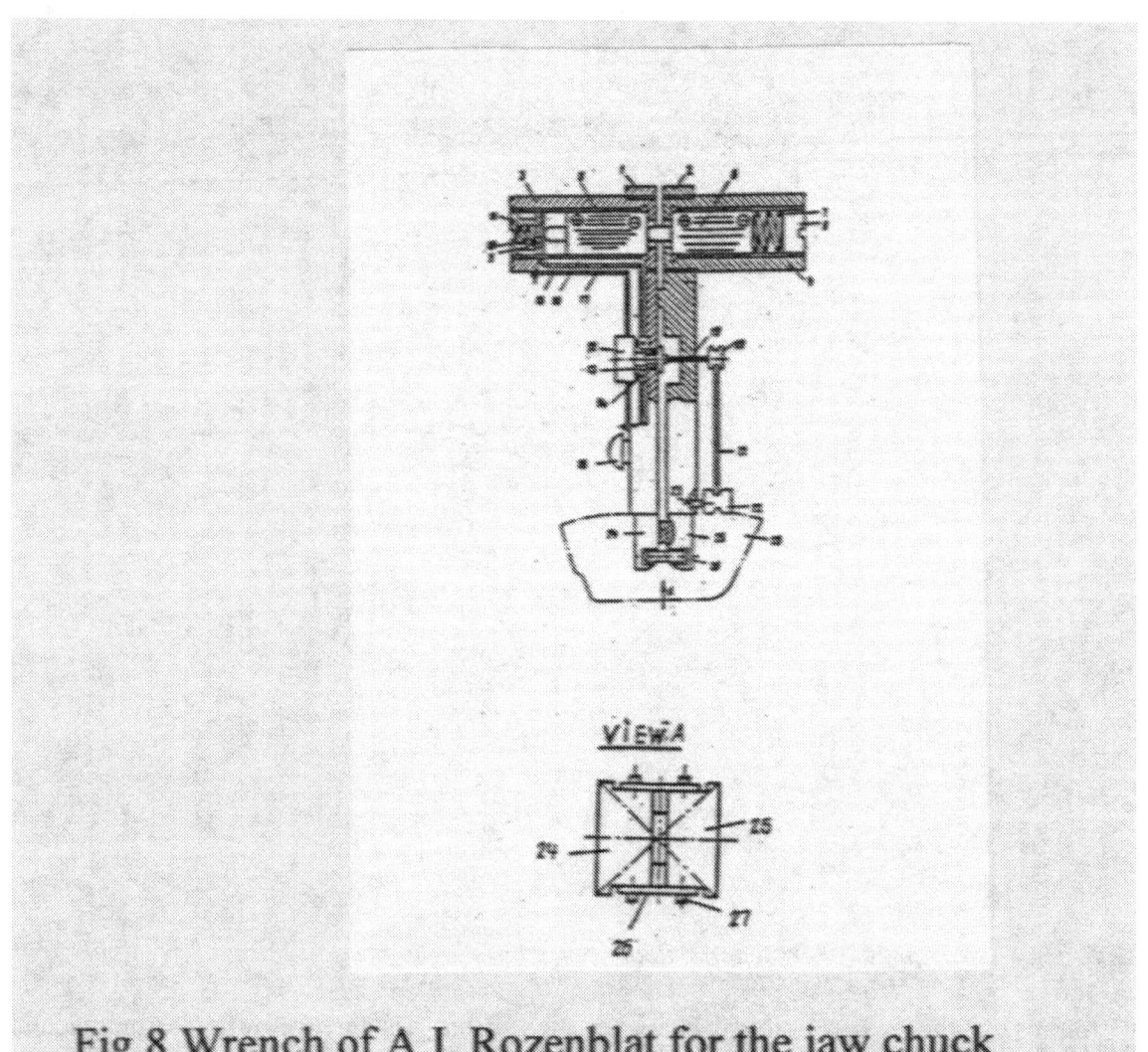

Fig.8 Wrench of A.I. Rozenblat for the jaw chuck

6. Progressive cutting tools and methods of machining work-piece with the increased allowances

The author established some facts by experimentally and later designed the new telescopical cutting tool, as shown in Figure 9.

This device can cut the very large irregularity allowances from the different forging work-pieces, such as crankshaft, rotor, etc.

The foregoing objectives can be accomplished by creating a telescopical cutting tool. The multiple insert is fixed so that the first cutting insert fits inside of the second insert which fits inside of the third insert. Each of the next inserts fit inside of each of the preceding ones so that each insert is movable and fixed on a separate holder.

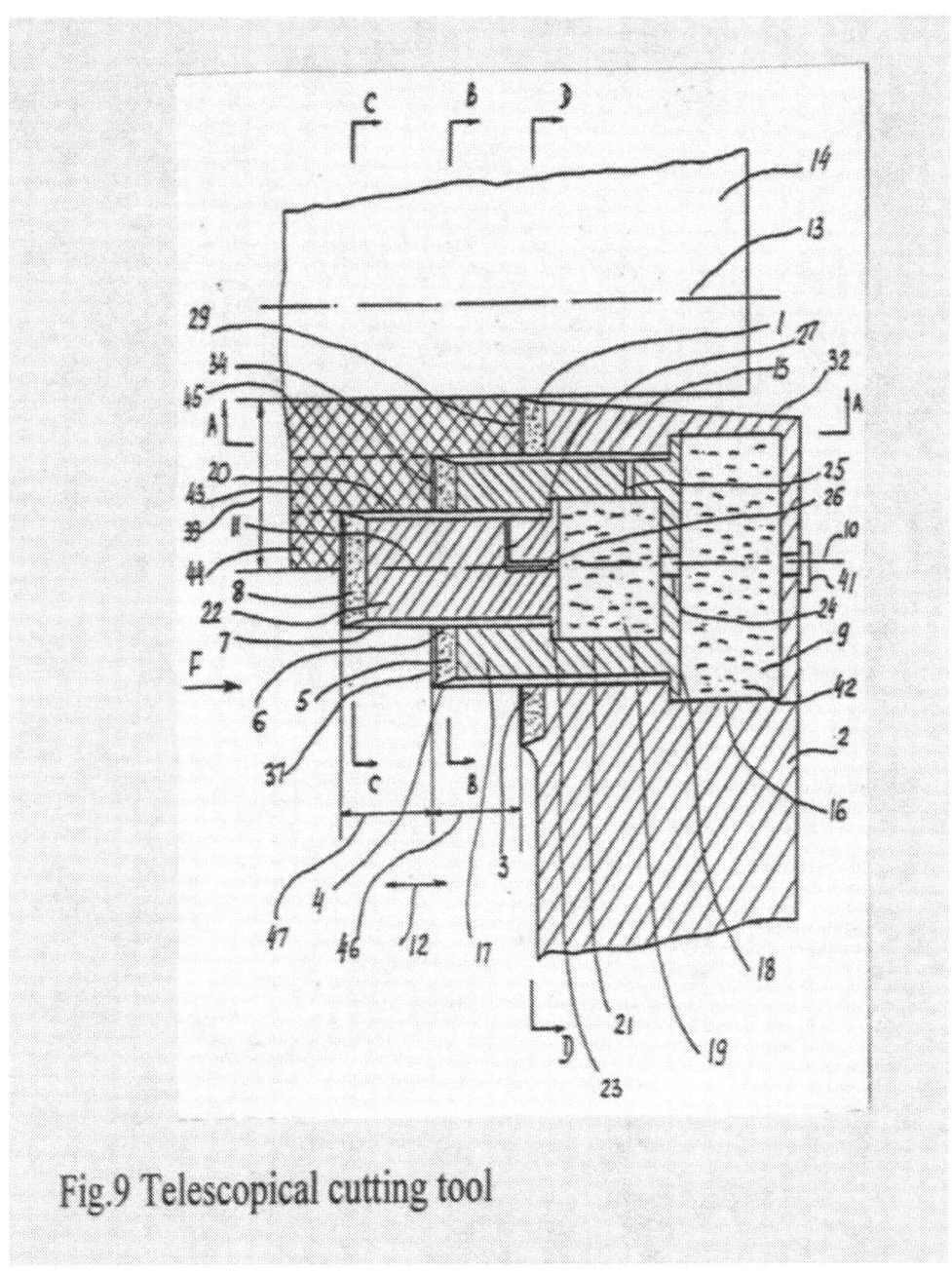

Fig.9 Telescopical cutting tool

- The other variety of new cutting tool which can turn the work-piece with large allowances, there is a multiple carbide insert, as shown in Figure 10.

Multiple carbide insert has general cutting edge is made in view of the interval line and in sum they combine on the perimeter the multiple insert.

Method of forming such multiple carbide insert is supplied by means that both inserts (the same sizes) fix on the axis of tool holder and then one of them turns relatively of the first insert, for instance on 45 degrees, and then projects the new contour of the multiple carbide insert on the frontal surface.

Multiple carbide insert permits to cut from work-pieces the large allowances because many cutting edges participate in the machining process and all the operation makes up in the one way of feed tool.

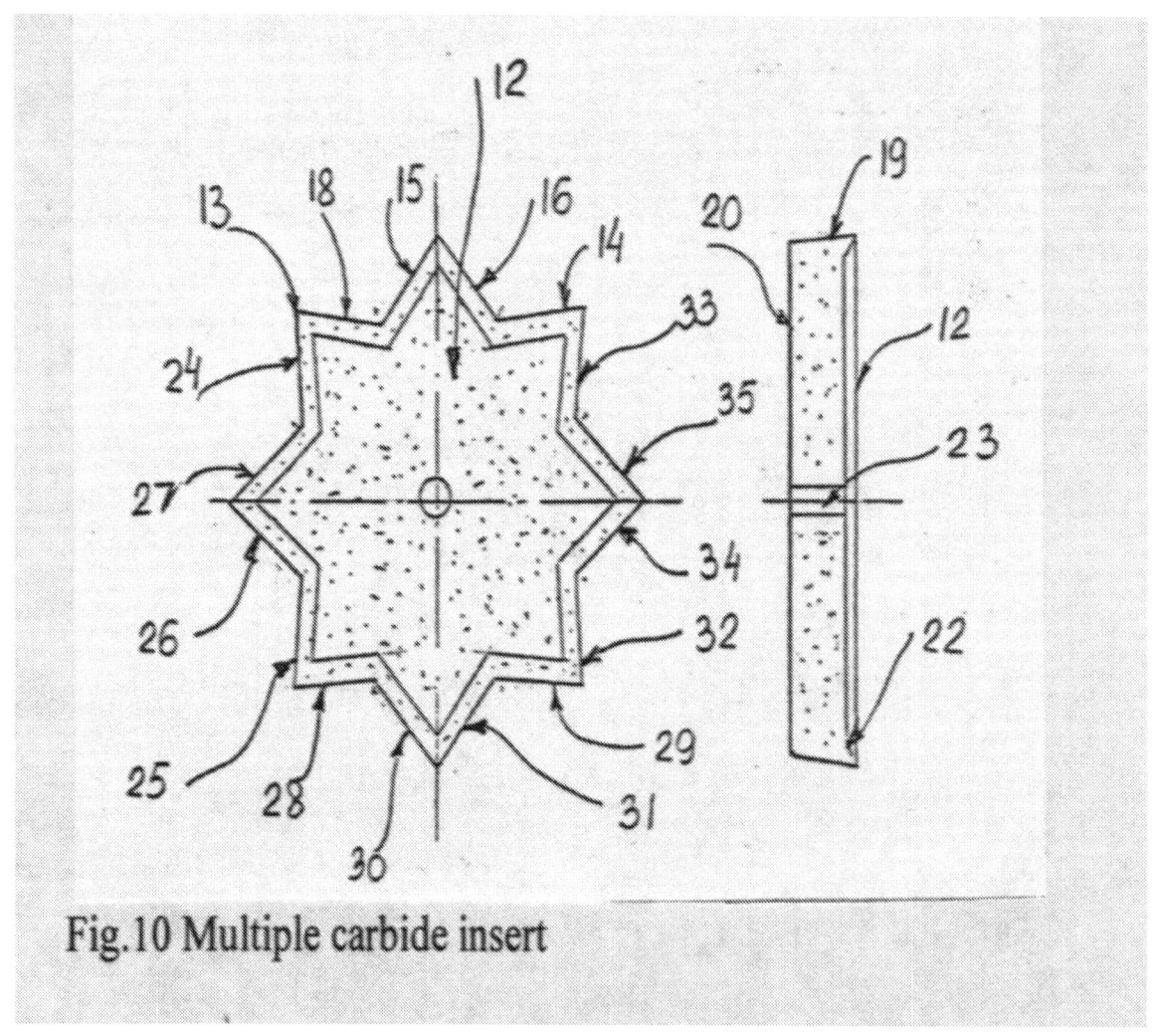

Fig.10 Multiple carbide insert

7. The improvements of break-chips in the cutting of stainless and heat-resistance steels

Changing of angles of the cutting tool in machine processing considerably improves the break-chips process, particularly for stainless and heat-resistance steels.

A variety of this cutting tool is the angular-variable assembling cutting tool of A.I.Rozenblat [12], as shown in Figure 11.

This tool is used to expand the technological possibilities by means of increasing the rigidity of the fastening of the cutting part of tool and reducing the wear in the machine processing of different materials, mainly the stainless and heat-resistance steels.

Materials cause continual wear and failure of the geometry of cutting tool, particularly on the face of a tool over which the chip slides, thus forming a chip breaker surface.

One of the benefits of this tool is to decrease wear and increase tool life during the cutting process.

This tool is comprised of a holder, a cutting part with a head, a stem installed in the holder on a support with the possibility of turning, and a turning mechanism interacting with the stem of the cutting part.

There is a support element installed on the head of the cutting part of the tool and is made in the form of two convex spherical plates interconnected by a pin with a spherical head.

Changing the angles of the tool makes it possible to decrease the wear of the tool, and also to break the chip that improves the cutting process in whole.

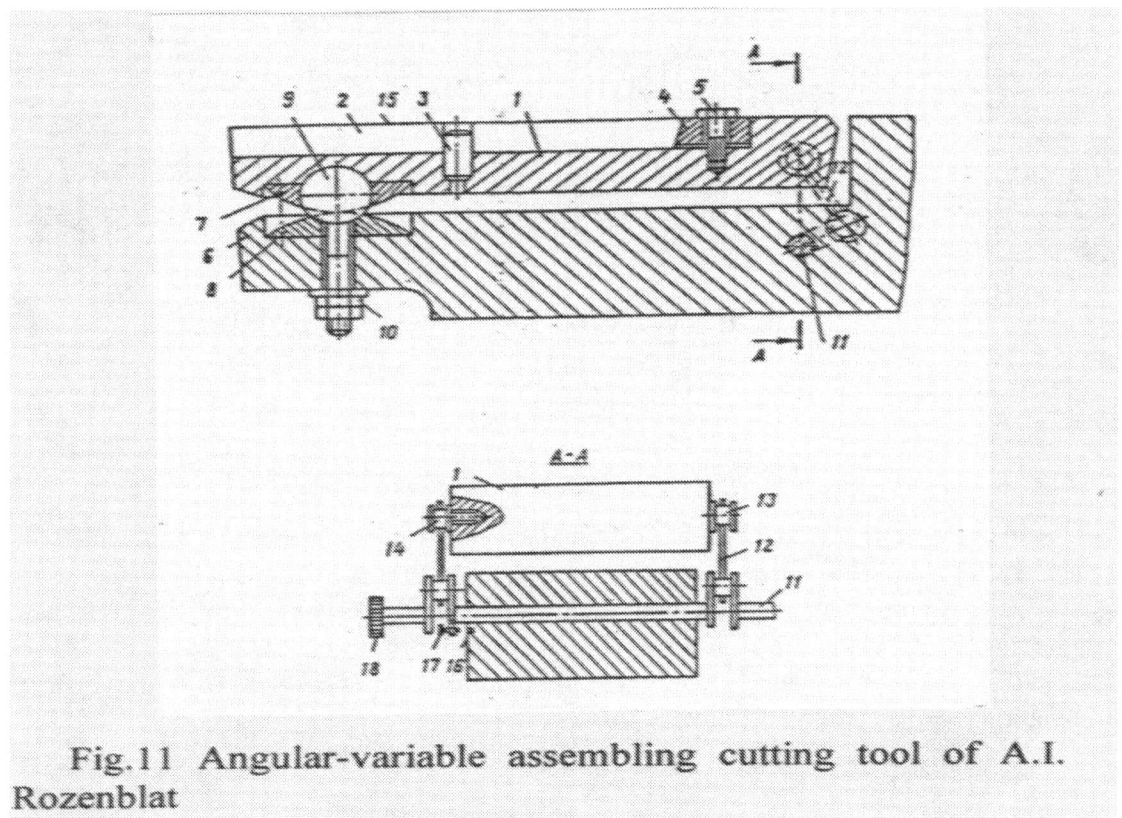

Fig.11 **Angular-variable assembling cutting tool of A.I. Rozenblat**

- The other variety of cutting tool for the improvement of break-chips is the break-chips insert, as shown in Figure 12.

167

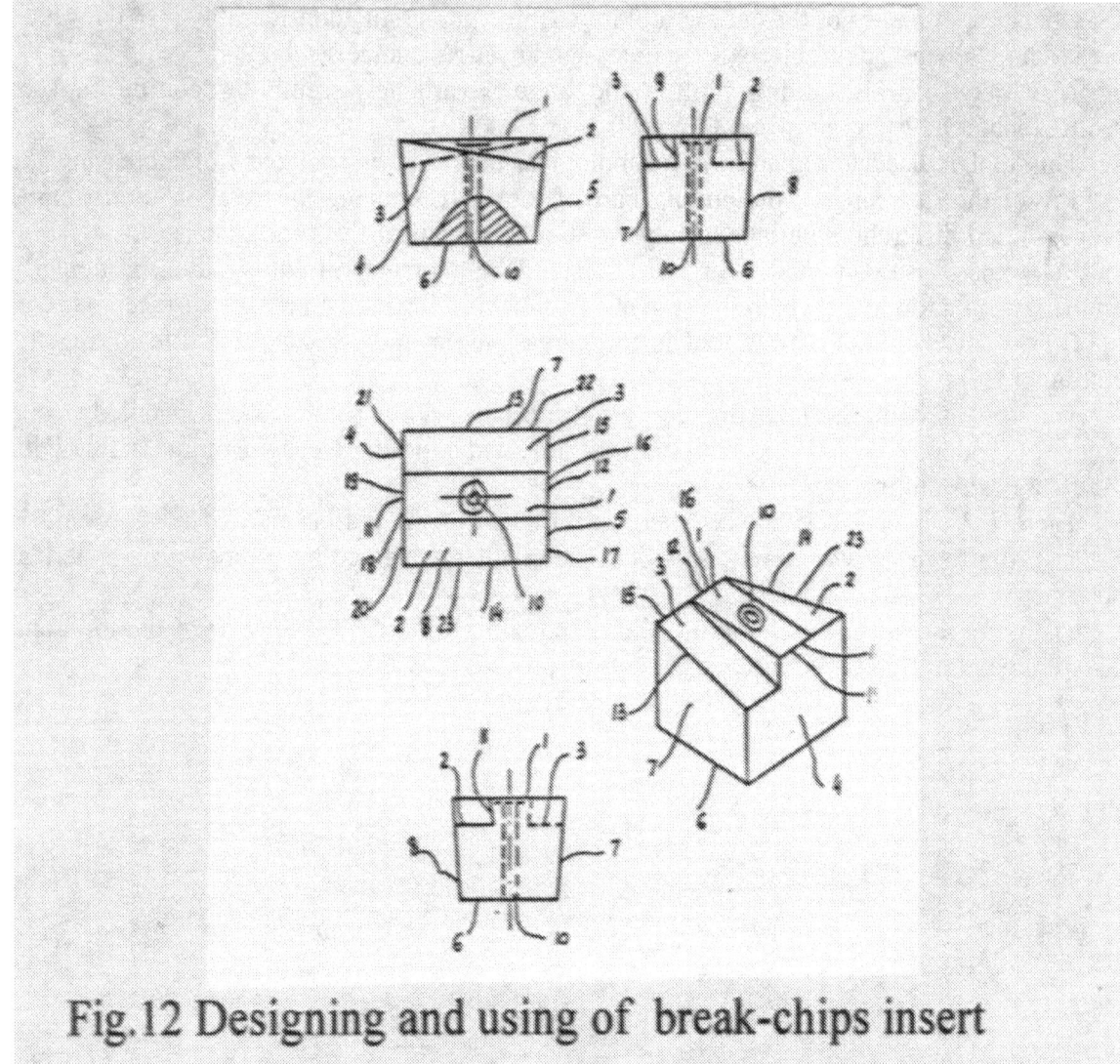

Fig. 12 Designing and using of break-chips insert

This insert is designed so that it has a frontal surface which is made with a variable rake.

This surface consists of three separate parts: the first part is made with the value of rake equal to zero, the second part made with the negative value of rake and the third part made with the positive value of rake.

Also the break-chips insert can have the variable front angles with the different combinations.

METHODOLOGY OF DESIGNING THE NEW CUTTING TOOL AND MACHINING ATTACHMENTS

Analyzing the above-named new cutting tools, the author thinks that it necessary to consider the main principles and methods of designing cutting tools and machining attachments for the manufacturing processes.

So, the following steps are most important in the design of tool:

1. Problem must be actual and practical;
2. The designer primary must do the technico-economical analysis of the problem;
3. Complex problems must be divided into simple problems and later to approach the designing in order of complication;
4. To sketch the designs of the multiple new decisions and by the methods of test and analysis to remove the useless technical decisions;
5. To investigate the suitable technical decisions again and to do the technico-economical analysis in conformance with the manufacturing processes where this tool will be used;
6. To investigate the patent and literature sources and to choose the prototype for the new cutting tool;
7. And finally, the designer must design the new cutting tool and to introduce the technico- economical analysis for presentation and discussion.

CONCLUSION AND RECOMMENDATIONS

1. The new cutting tools and methods suggested by A.I.Rozenblat are original and actual for the manufacturing industry, but demand the careful analysis and research before using their in large-lot production;

2. The author thinks that present guiding materials are not perfect for the designer because they do not have enough information and data, particularly in the area of cutting tools and progressive dies, and also in pneumatic transportation of chips from machining processes.

3. Also the author recommends the publishing through the ASME or SME, some quarterly material for the designer about: *"Express information Bulletin. Cutting Tools"* which can cover a wide range of manufacturing questions.

REFERENCES

[1] H.W. Pollack, Tool Design.—Prentice Hall, Englewood Cliffs, NJ (1988).

[2] R.B. Aronson, "Machine Tools of the Future", Manufacturing Engineering. - SME, Vol.113, Number 1,1994, pp.39-45.

[3] Mark's Standard Handbook for Mechanical Engineers, 9th Ed., pp.13-68.- McGraw Hill Book Company (1987).

[4] Machinery's Handbook. Twenty-fourth Ed., pp.686-691.- Industrial Press Inc. (1992).

[5] Regal Industrial Sales Co., Inc. 1985 "Catalog No.85 ", Bensenville, Illinois, pp.130-131.

[6] Author's Certificate # 360155.USSR.MPC B 23, B 29/04. Cutting Tool / Rozenblat A.I./USSR/. # 1364370/25-8. Applied on 8-15-69. Published on 11-28-72, Bulletin #36. "Discoveries. Inventions".1972.#36, p.31.

[7] Author's Certificate # 1504064.USSR, MKI B 23 Q 41/04, B21 J 7/00. Billet Processing Installation of A.I. Rozenblat. Rozenblat A.I. USSR #4305946/25-27. Applied on 9-14-87. Published. Bulletin #32-1989. "Discoveries. Inventions".1989,#32.

[8] Author's Certificate #1131634. USSR, MKI B23Q 11/02. Tool Cleaning Device. Rozenblat A.I. USSR #3435998/25-8. Applied on 5-10-82. Published on 12-30-84, Bulletin #48. "Discoveries. Inventions". 1984. #48, p.35.

[9] Author's Certificate #1199466. USSR, MKI B23b 27/04. Assembling Cutting Tool of A.I.Rozenblat. Rozenblat A.I. USSR #3766852/25-08. Applied on 7-6-84. Published on 12-23-85. Bulletin #47 "Discoveries. Inventions".1985.#47, p.53.

[10] Author's Certificate #1500416. USSR, MKI B21 87/08, B21 28/14. Multi-operational Combined Die of A.I.Rozenblat for processing Sheet Materials. Rozenblat A.I.USSR.#4354771/25-27. Applied on 12-8-87. Published. Bulletin #30-1989. "Discoveries. Inventions".1989.#30.

[11] Author's Certificate #1505676. USSR, MKI B23b 31/00. Wrench of A.I.Rozenblat for lathe Chuck. Rozenblat A.I. USSR #4291302/25-08. Applied on 7-28-87. Published, Bulletin #33-1989. "Discoveries. Inventions". 1989,#33.

[12] Author's Certificate #1199466. USSR, MKI B23b 27/04. Assembling Cutting Tool of A.I.Rozenblat. Rozenblat A.I. USSR #3766852/25-08. Applied on 7-6-84. Published on 12-23-85. Bulletin #47. "Discoveries. Inventions". 1985,#47, p.53.

2. RUSSIAN INVENTOR AND SCIENTIST BRINGS THE NEW TECHNOLOGY TO USA

ABSTRACT

This paper will give a brief review of inventions in the field of Mechanical Engineering with special attention give to the high technology processes for commercial practice and further research in the different manufacturing processes in the 21st Century.

Specifically, there are four types of inventions discussed here: environmental, safety of flying, advanced machining processes and cutting tools, the efficiency of some machines and mechanisms is also discussed.

In addition, this paper indicates basic design and research activities in Mechanical Engineering in 1998 in the United States.

A. ENVIRONMENT PROTECTION

1. Designing vehicles with zero emissions

The main purpose of this effort is to review and reevaluate some innovations and advanced methods in the patents and literature, in order to discover a way of solving some problems regarding emissions, and also to clear the way for future research in this field.

As our population increases and industrialization becomes more intensive, the problem of pollution becomes more serious in today's world [1].

The study is done by Nurnberg [2] shows that automobile exhaust gases are the main cause of pollution in the streets, particularly during the day time when the intensity of transportation increases.

A recent article [3] described a bus with minimal exhaust emissions that used liquid hydrogen fuel.

However, this device demands the big capital investments in the designing and production processes.

As is generally shown in references [4], [5], [6], [7] and [8], there are many other systems for gathering and storing exhaust gases in special tanks on vehicles (boats, aircraft's, etc.), but these methods and devices do not improve the environment as a whole.

The requirements of designing a new vehicle with zero emissions to the environment are:

- Expenses must minimal;
- Gasoline or diesel fuel must be used for the vehicle because it is the cheapest source of energy;
- The vehicle must the zero emissions to the environment.

This paper describes a new model of automobile that meets the above-named requirements, as shown in Figure 1.

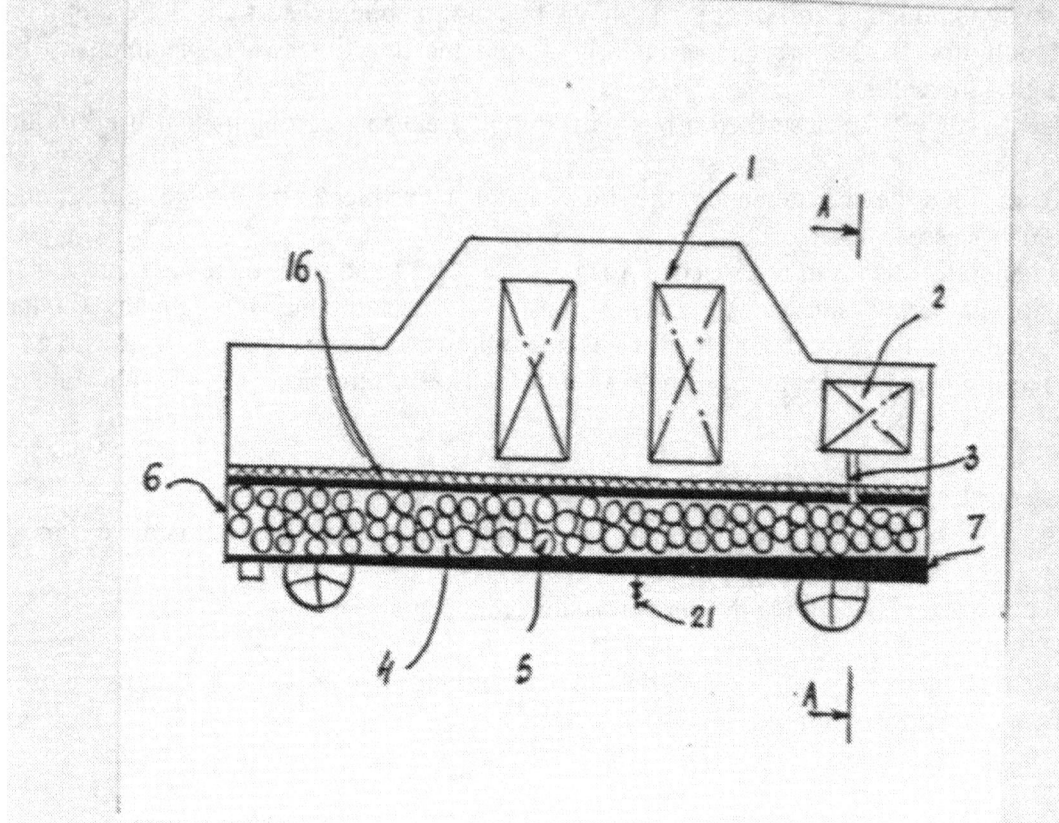

Fig. 1 Combined automobile with zero emission

The combined automobile has an exhaust pipe which is connected on one side with the source of the exhaust gases, and on other side with a vessel which is closed hermetically to the environment.

And besides this vessel is made in form of a rectangle and has an inclination on the lower inner surface of it. Also, this vessel on both sides has covers which are joined firmly and has the pistons with the coils moving axially into its surface.

In inner space of said vessel then inputs the spherical balls everyone of each has the spring-and-balls valves (non-returned).

During the working process of the engine, the balls absorb the exhaust gases. After they are filled, they are unloaded from the vessel of this automobile to the special installation located on special storage station.

B. SAFETY OF FLYING

1. Helicopter

The present paper provides a very superficial introduction to the designing of well-known helicopters and relationships of it to the safety of flying.

For those who seeking more information the Encyclopedia [**9**] and innovations [**10**], [**11**] provides an encyclopediacal review of this field. As the USA continues its leadership in manufacturing, innovations in high technology, particularly in the design of the new helicopters and aircraft, are very important to maintain their leadership for the future.

This section describes a new helicopter design, as shown in Figure 2, which guarantees the safety of flying.

This innovation utilizes the construction in view of n-quantity screws located on the vertical and parallel axis of rotation which have movement from the transmission system.

The diameter of each screw in whole determines by experimentally with given distance and safety conditions.

Also, each screw has a movable axis of rotation which makes the back-progressive movement in the vertical plane by means of hydraulic system.

In the process of flying the auxiliary attached screws produce more summary aerodynamic flow which increase the speed of flight and improves the maneuverability of the helicopter as a whole.

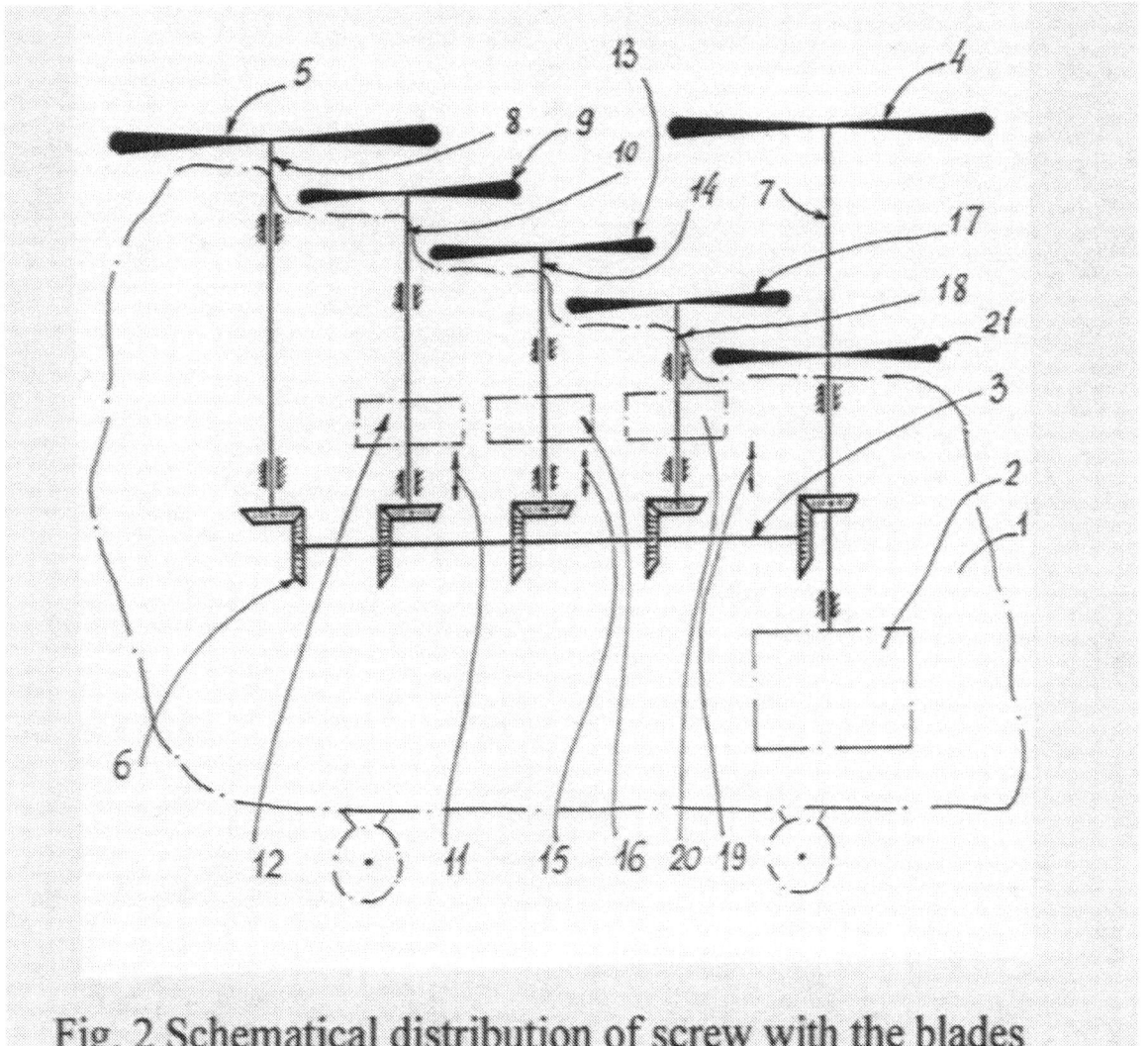

Fig. 2 Schematical distribution of screw with the blades

2. Aircraft-passenger

This innovations, as shown in Figure 3, improves the safety of flying, particularly in emergency situations (engine broken, fire, etc.).

The new element of this aircraft there is the fuselage which is made with a movable inner element and besides it has the electromagnets which are installed on both lateral sides. They are joined with the front and tail parts of the aircraft which have opposite terminals.

Also the method of flying includes such operations as the primary separation from the tail part and the removal of the movable outer element of fuselage with the fixed wings, and finally the movable inner element which has electromagnets installed on both lateral sides.

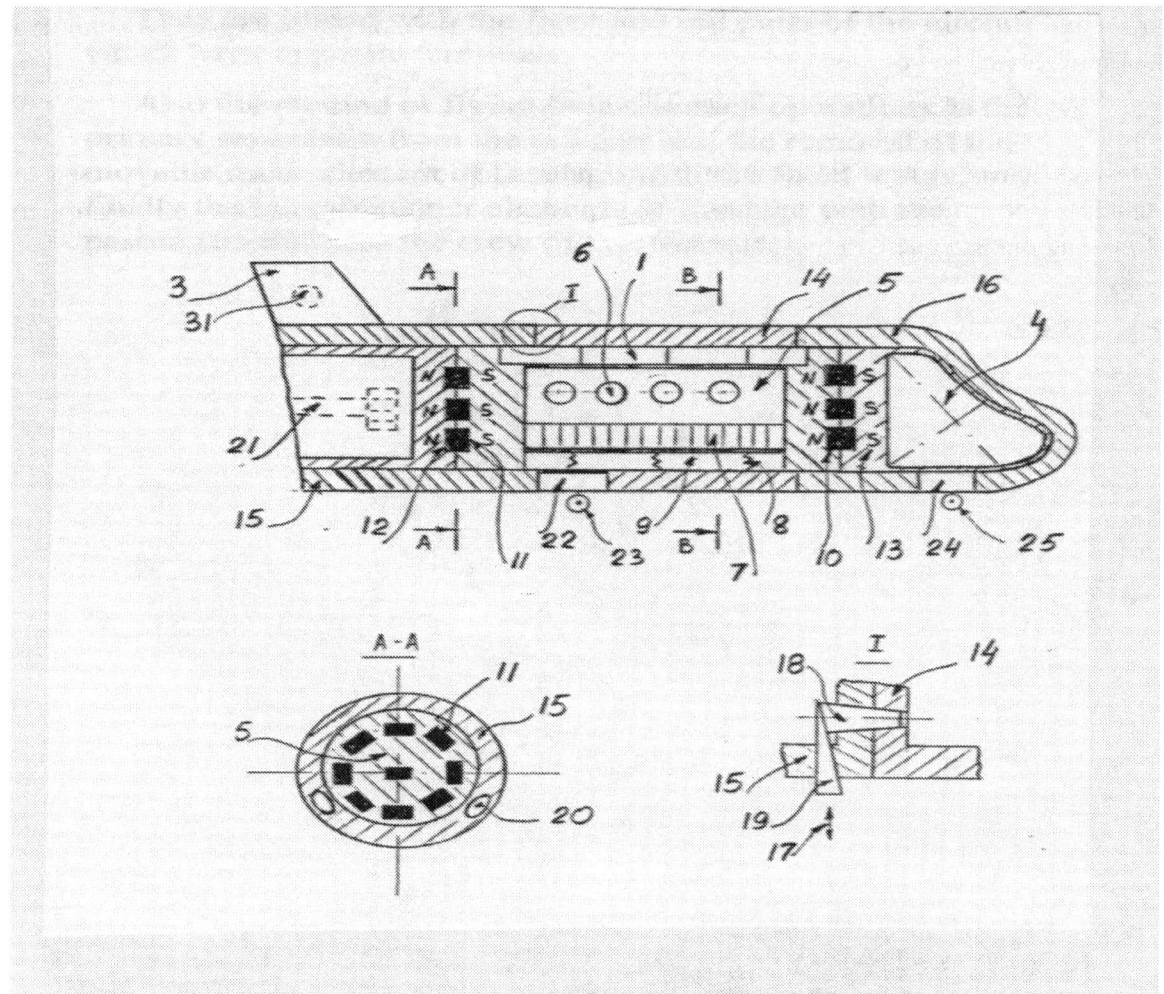

Fig. 3 Aircraft-passenger with the moveable elements

C. ADVANCED MACHINING PROCESSES

1. Method of processing curvilinear channels

This method is related to the mechanical processing of curvilinear channels predominantly in elbow of systems with pneumatic and trunk lines transporting the ferromagnetic or abrasive materials [12].

It is a movable device, as shown in Figure 4, supplied with a compound of moving, separate parts, assembled elbows and working tools for processing of these elbows.

Accordingly with the invention the elbows and tools have the rotary motion.

The new element is that working tools are filled into the curvilinear channel of assembled elbows, so that the end of a channel of each preceding tap coincides with beginning of a channel, and together they form a closed curvilinear space for the tools. This is not difficult device.

The method of processing requires that the concavities of all taps are turned toward the center. The working tools are charged, and all system has rotary movement.

This is a modern invention which means that the assembled system has dynamic and complex characters of movements.

Also centrifugal forces press the balls to the surface of the channels and they start to move toward the center.

This makes it possible to intensify the processing of elbows. This device is used to increase the efficiency of the mechanical processing of curvilinear channels mainly to finishing out and decrease the abrasion.

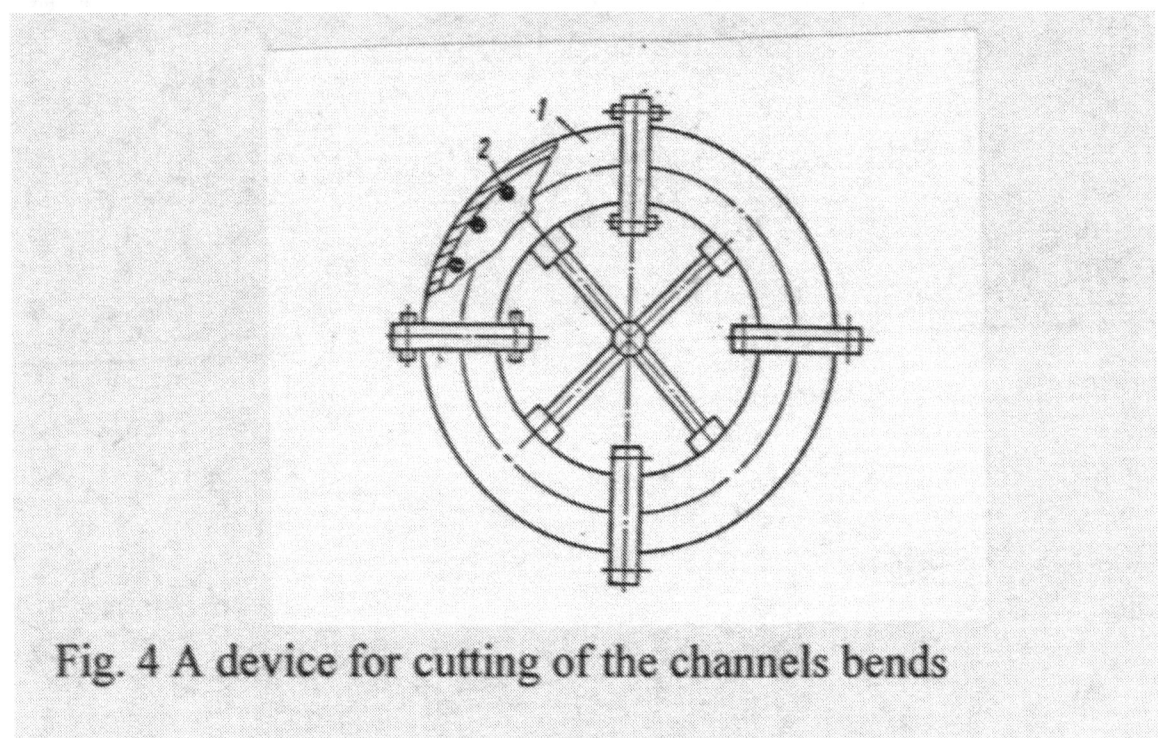

Fig. 4 A device for cutting of the channels bends

2. Magnetic-dynamical pipeline

This device is used to decrease abrasion of elbows and pipeline surfaces during the transportation of ferromagnetic materials.

Pipeline are used for pneumatic transportation of a variety of materials, including metallic particles, such as iron or steel either as part of an industrial process or as a method of waste accumulation and processing. Such metal particles have a very high abrasion index, causing continual wear and damage to the pipeline which transports it.

This damage is most pronounced at the elbow and curves of the pipeline when the differential inertia of the particles sends them against the pipeline wall.

This invention, as shown in Figure 5, utilizes a non-metallic element of the pipeline, combined with the magnetic field created by magnets of the same pole which are situated opposite of one another.

These magnets create a magnetic field which present repellent forces so that the ferromagnetic material does not touch the inner surfaces of the elbow and pipeline, reducing contact with surfaces and the resulting abrasion.

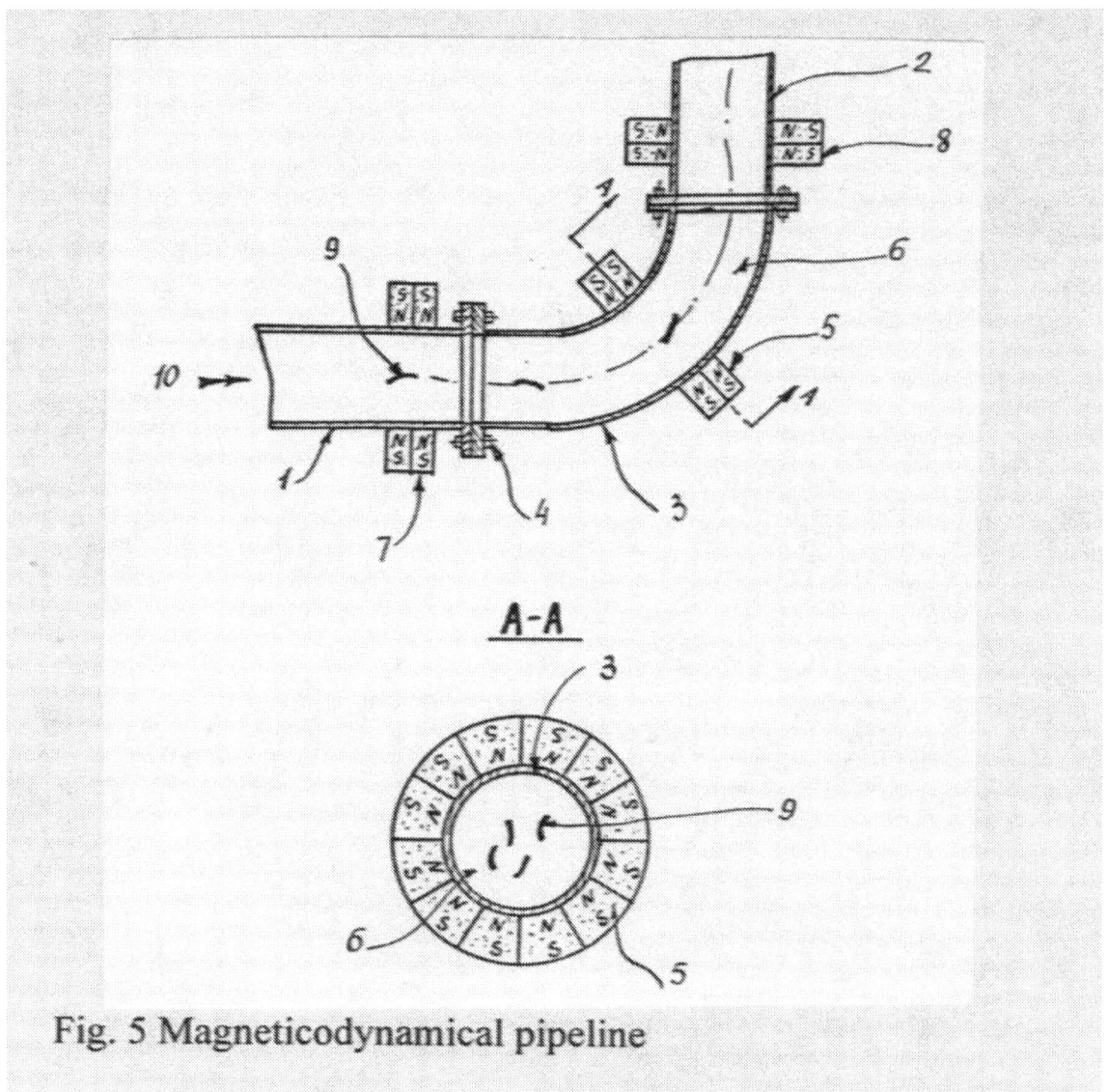

Fig. 5 Magneticodynamical pipeline

3. *Wear-resistant pipeline elbow*

This invention is related to the pipeline technology for the pneumatic transport of loose materials.

The aim of this device [13] to increase wear- resistance due to eliminating wear from the bend walls. As shown in Figure 6, the bend is made from the non-magnetic material.

Rolling-contact bearings are fastened diametrically on the body, a yoke with magnets is installed on the bearings. The yoke is set into rotation around the bend generating a rotating magnetic field inside of the latter.

The ferromagnetic material moves inside the bend over the spiral-shaped trajectory revolving uniformly around the internal walls.

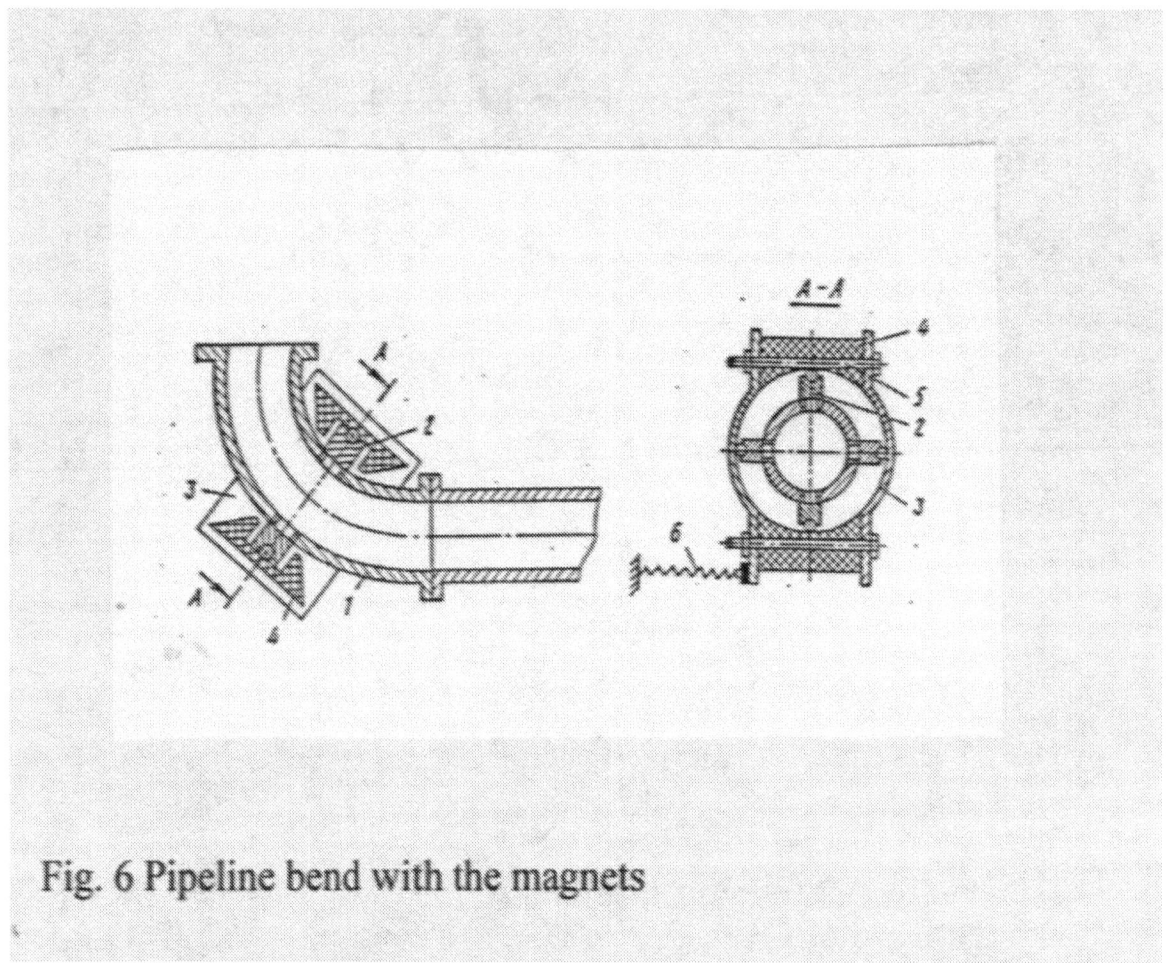

Fig. 6 Pipeline bend with the magnets

4. Pneumatic installation

Pneumatic installation is used for removing ferromagnetic chips and iron dust from the cutting zone of machine-tools by means of a pneumatic method. In Figure 7 is shown schematically the pneumatic system in whole.

The other important objective of this device is to break chip, particularly the stainless and heat-resistance steels which form in the process of cutting these materials. This chip causes many problems for the machinist and machine-tool user.

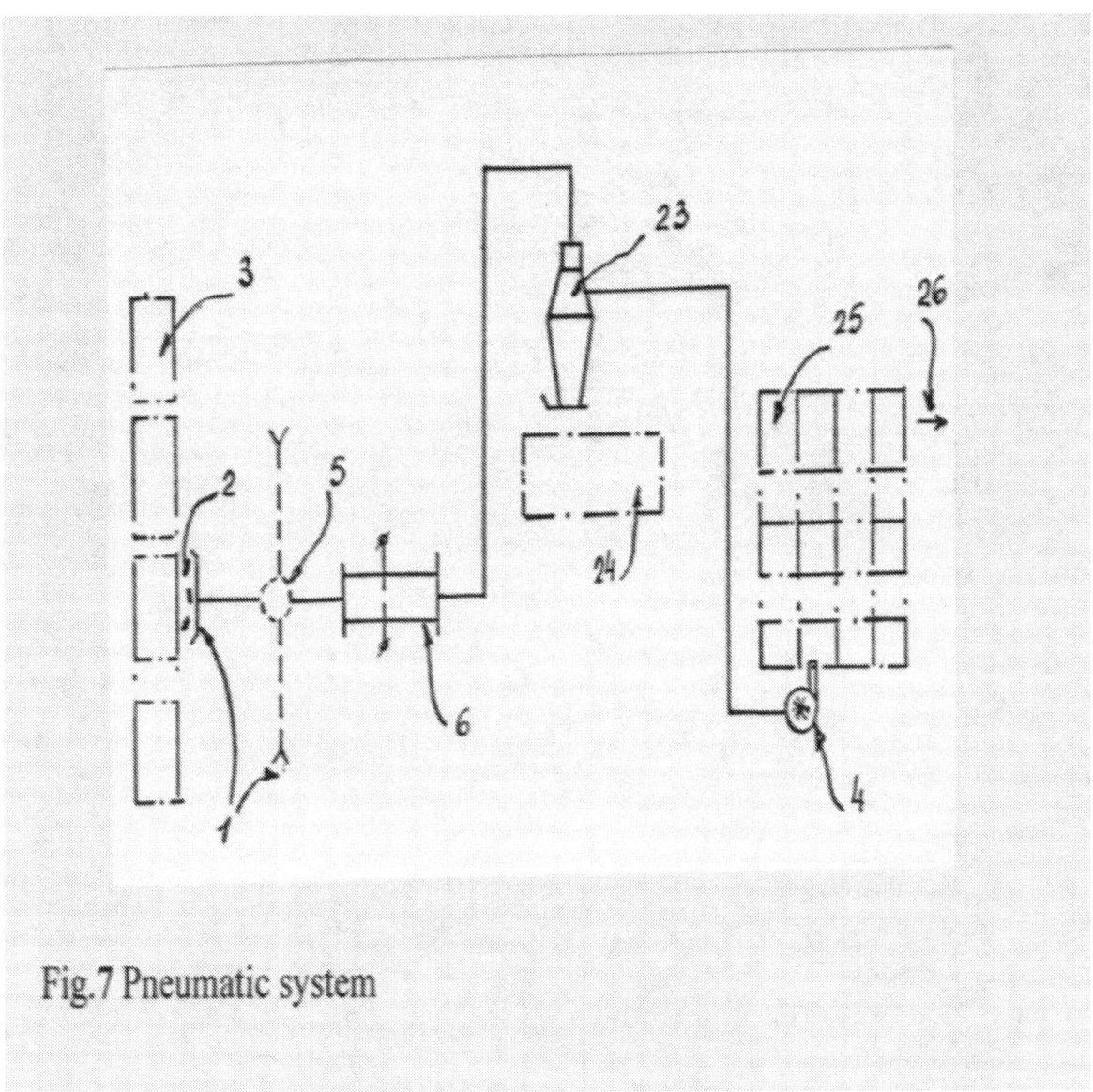

Fig.7 Pneumatic system

In pneumatic systems of pipeline, the plasmotron fixes between the dust-chip arrangement and cyclone which has two carbon rods, supporting in process of transportation ferromagnetic material the high voltage arc as shown in Figure 8.

The plasmotron also has a movable element in view of plain matrix having many holes so that the frontal surface of this matrix is made concave and in the lower part of plasmotron, on the way of movement of material, is installed the water system.

In the process of cutting material the chip comes into the dust-chip arrangement and then comes into the zone of arc of the plasmotron. In this place the chip melts and further moves through the holes of matrix, forming the granules (spheres) by means of cooling from the water system.

By that, in zone of plasmotron the chip receives a form of granule (convention named as sphere) and naturally in this period the coefficient of friction changes as the element chip (friction of slipping) converts into the sphere, i.e there is the forces of friction rolling.

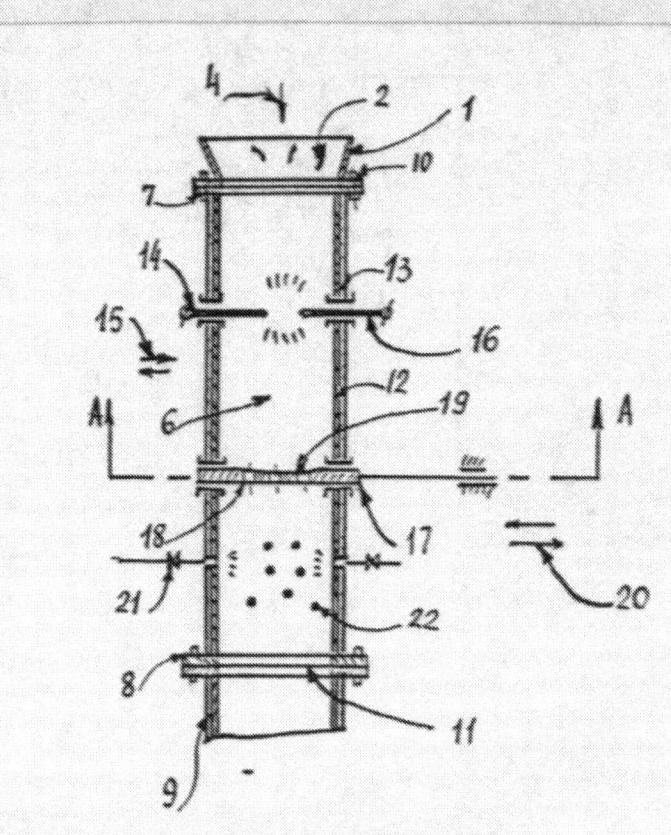

Fig. 8 Plasmotron in pneumatic systems

5. Pipeline for pneumatic transportation of granular materials

Pipeline for pneumatic transportation of granular materials used, particularly for transportation of different granular abrasive materials such as sand, ash and metallic particles (iron, steel, bronze, etc.), either as part of an industrial process has a short period of durability for the pipeline, particularly of the elbow.

A suggested device, as shown in Figure 9, is used to decrease the abrasion of the elbow. The fixture for twisting of movable granular material is installed between the straight part and elbow pipes of pneumatic systems.

This fixture is equal in size as the same pipe through which it passes and arranged with the possibility of rotation from a drive so that its axis of rotation is displaced along the longitudinal axis of branch pipe [14].

This method of pneumatic transportation gives also the possibility of decreasing losses of pressure and power consumption during the transportation of granular materials.

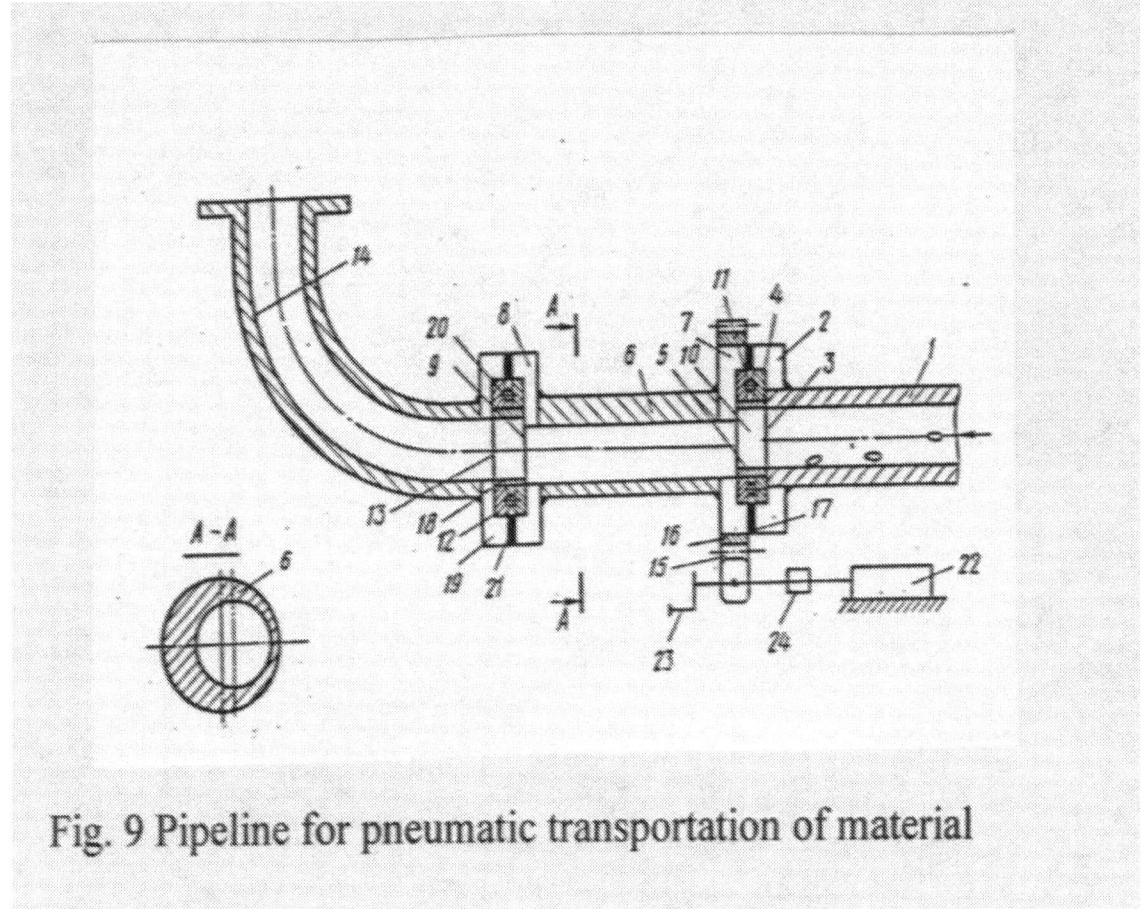

Fig. 9 Pipeline for pneumatic transportation of material

6. Internally-modifiable pipeline transport

These pipelines are used for transporting a variety of materials, including solid materials in a liquid suspension. Such materials cause continual wears and damage to the pipeline which transports it due to its constant contact with the inner surface of the pipe.

This device [15], as shown in Figure 10, utilizes a uniquely designed insert at such bends and elbows.

This insert alters the trajectory the suspension as it enters the curve, converting it to a more homogenous stream with reduced separation of the abrasive materials. This modified trajectory has been shown to materially reduce the abrasion to the pipeline.

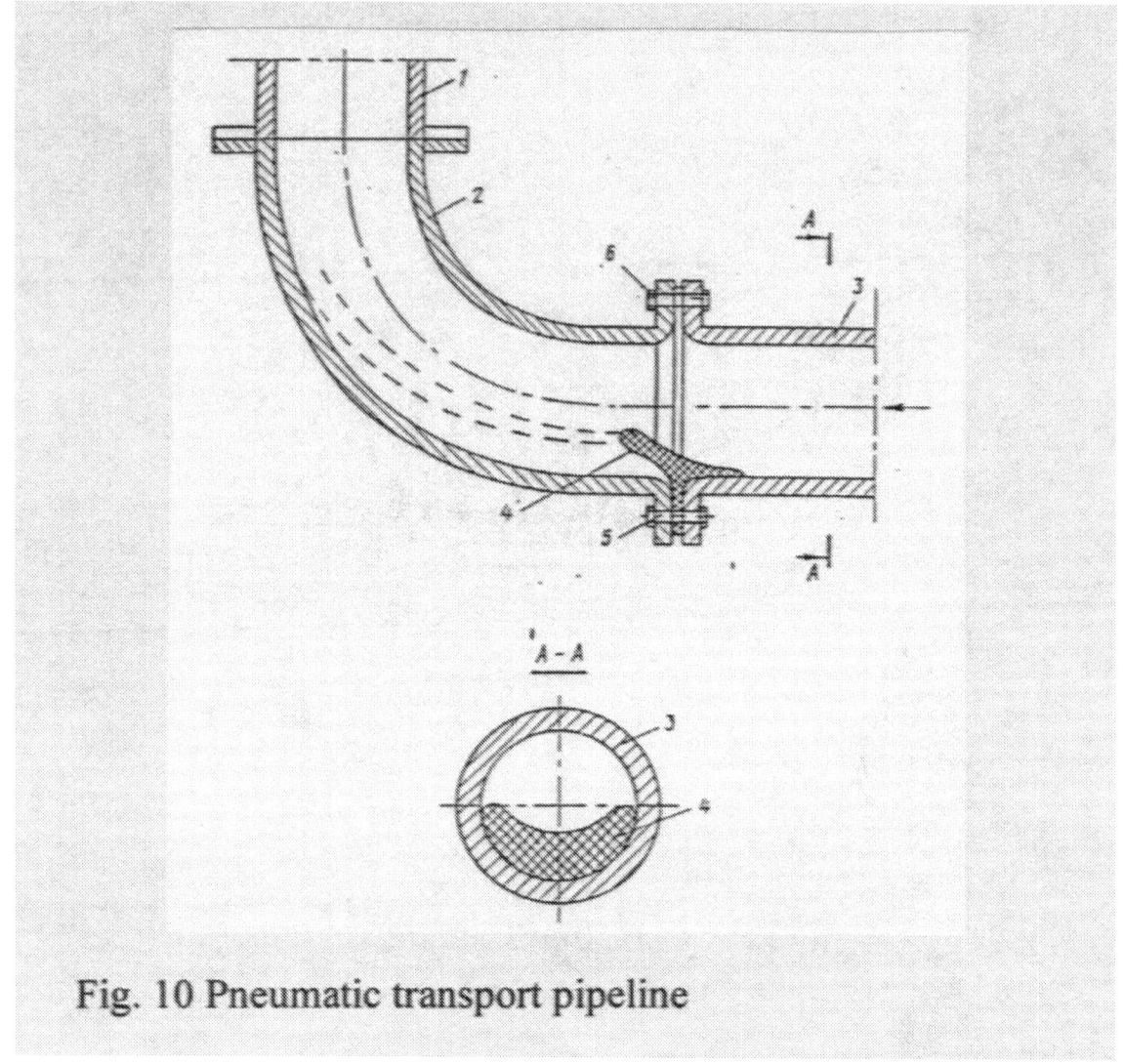

Fig. 10 Pneumatic transport pipeline

D. EFFECTIVENESS OF WORK SOME DIFFERENT MACHINES AND MECHANISMS

1. Combined engine

The fundamental prerogative of the present device proves a high effectiveness of coefficient useful action (c.u.a) in comparison with well-known engines (value of c.u.a composes 0.30).

The other important objective of this device is the reduction of material-capacity the complex power installation (engine, transmission, system of electricity, etc.), mainly for the maritime ships and submarines.

In this device, as shown in Figure 11, the piston has the permanent or electric-magnets which are rigidly mounted and fixed on the periphery of lower part of said piston.

The cylinder of this engine has the elements of coils including one or more separated coils which are connected in a series circuit and installed inside of said cylinder on the perimeter, along of its surface.

In the process of committing a full working cycle by the engine the piston, with magnets (they fulfill the function of a solenoid), produce the magnetic field which by the Farady's law, in a winding of excitation (they fulfill a function of induction—coil) induce an electromotive force (e.m.f) and additional current.

So, this device improves the parameters of a vessel in whole because the coefficient of useful action of the engine increases.

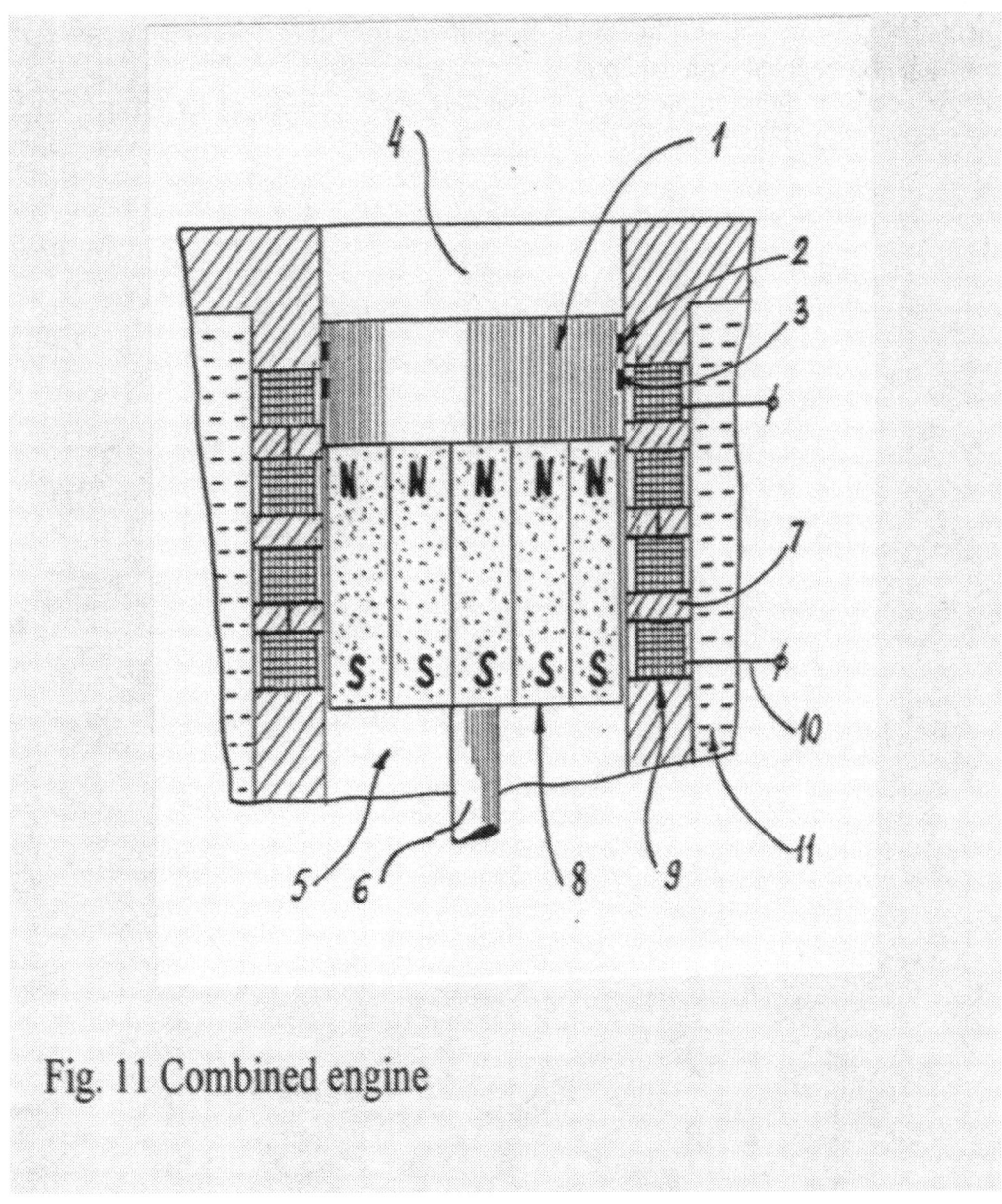

Fig. 11 Combined engine

2. Seismic combined nuclear system

This device is used for nuclear power plants and industry. The present well-known nuclear systems [16], [17] and [18] do not guarantee the safety of nuclear power stations as they are situated on the solid ground and exposed to seismic destruction.

Also the pressurized water reactors (PWR) are situated firmly and vertically, which demands the high buildings for this objective.

These defects are removed by this device, as shown in Figure 12, which is used for increasing the security of the nuclear system in whole.

This nuclear system consists of a reactor vessel with core, a steam generator, a main coolant pump, a pipeline and also other elements which are joined in one energetic system.

The new elements of this device are that all system is made horizontally and arranged in view of cylinder and piston, so that the function of said movable piston makes up the reactor vessel with core, embracing on this circle by combined cylinder, having the inner bimetal surface which is connected with said piston.

This cylinder has the multiple exhaust channels on peripheral of its surface for joining with cavities of said cylinder and other elements for movements of coolant water in working processes.

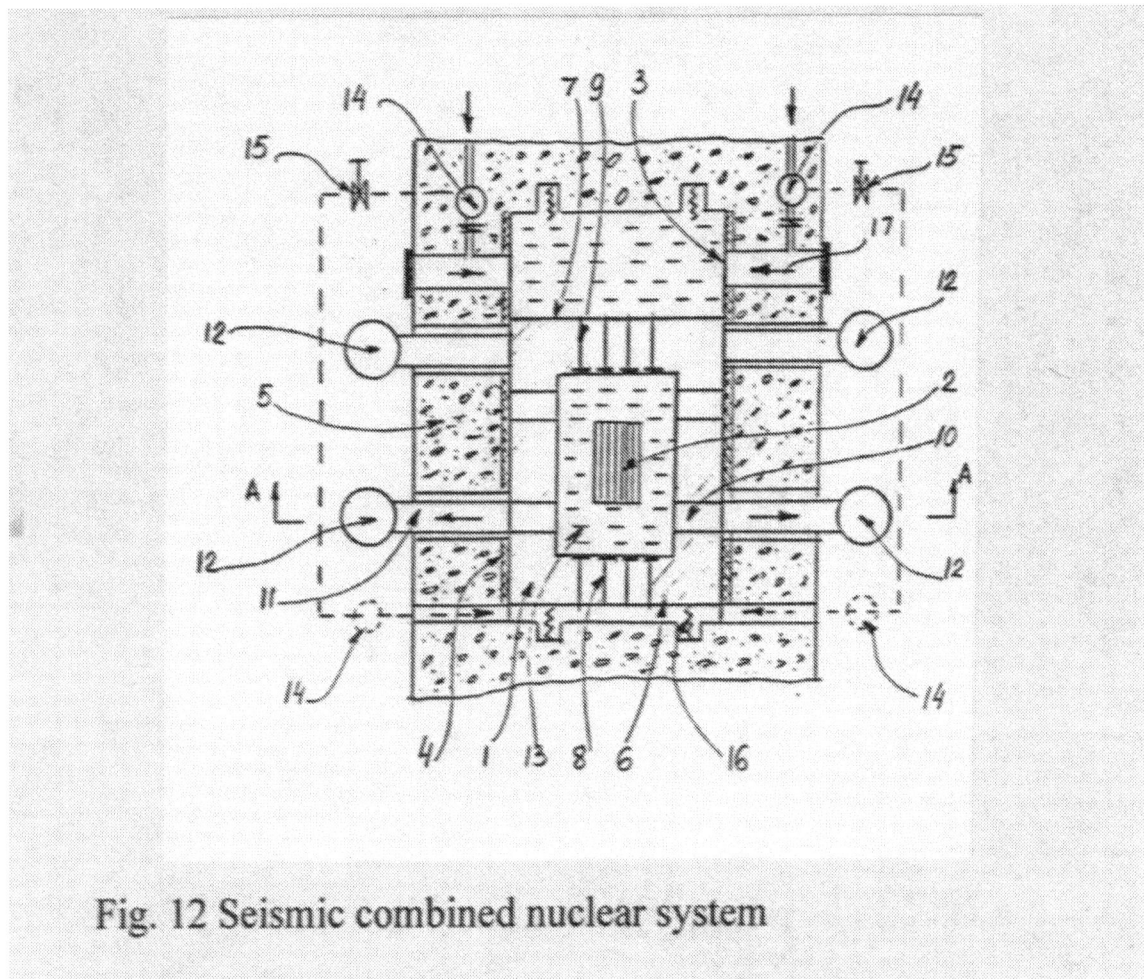

Fig. 12 Seismic combined nuclear system

3. Combined rocket "AIR"

The present device relates to space and rocketry techniques mainly to liquid propellant rockets and used for reducing the metal capacity of system and fuel economical components of rockets in whole.

Well-known rockets [19], [20] and [21] have a large weight and low coefficient of useful action. This device, as shown n Figure 13, is used to remove these defects.

The new element of this combined rocket is that the body made in view of inner motionless part which is connected through support bearings with a movable outer element having the aircrew blades, so that they fixed only at one side and displaced in according one to the other.

By the present of the movable element create additional the power system. Also these stabilizes on rocket are situated underneath of the exhaust system that make this device more useful for application.

The engine is joined with this additional power system so that movable outer element uses the exhaust gases from engine and rotates the aircrew blades of this combined rocket.

This device guarantees economy of fuel components and also decreases the weight of rocket in whole.

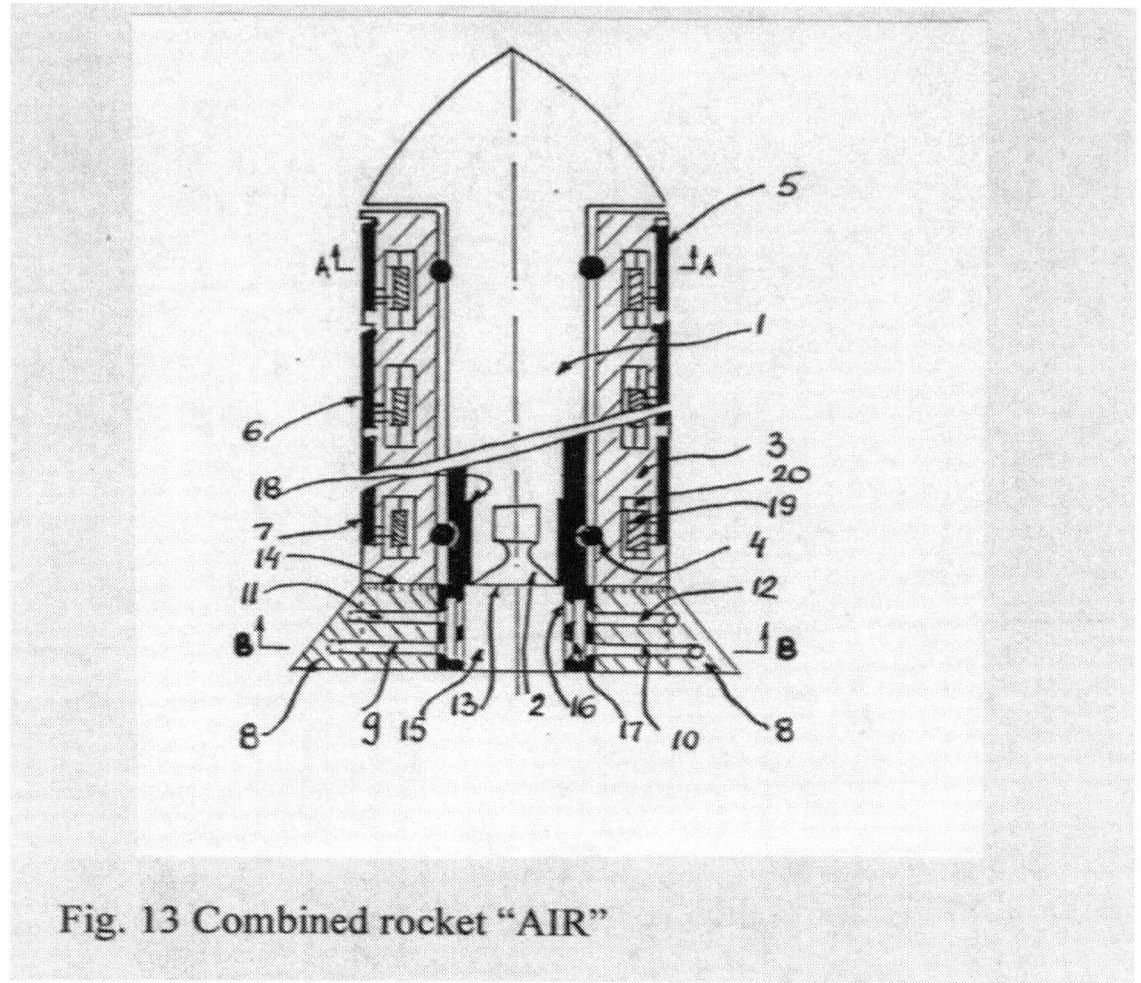

Fig. 13 Combined rocket "AIR"

E. CONCLUDING REMARKS

1. The innovations are suggested by the author destine advantageously for commercialization and demand the next evaluation and research works for application in industry.

2. At the same time the author would like to have priority, by publishing these innovations at this book for the 21 st Century, and to fix these creative works, which are shown here, by Copyright

REFERENCES

[1]   Kenneth, M. The biology of Pollution (1979),2nd Ed., University Park Press, Baltimore, pp.1-18.

[2] Nurnberg, H.W., Pollutants and their Ecotoxicological Significance (1985).

[3]   Винокур Г, (1997) "Безвредный автобус ".- American-Russian weekly Newspaper" 7 days, p.3.

[4]   Valenti, M. (1996), Bagging Car Exhausts.-J. Mechanical Engineering, ASME, vol.26,, p.26.

[**5**] West Germany Patent #2040640 K1 14k,3/08,1972. Motor Vehicle Exhaust Anti-pollution Device.

[**6**] West Germany Patent # 2440723,1974 Exhaust Gas Storage for Collection/Disposal on Central Station.

[**7**] U.S Patent # 3,065,774 Cl.141/38,1962 Device for Inflating Objects.

[**8**] U.S Patent 4,731,992 Cl. 60/281,1988 Fuel Supply Device for Engines of Aircraft.

[**9**] The New Illustrated Science and Invention Encyclopedia (1989).

[**10**] U.S Patent 4,531,692 Cl. 244-17.19,1985
Helicopter Flight Control and Transmission System

[**11**] U.S Patent 4,899,957 Cl.244,1988
Helicopter with Auxiliary Prolusion

[**12**] Author's Certificate # 1441658 USSR, MKI B24, B 39/02.
A.I.Rozenblat's Method for Processing Curvilinear Channels. Rozenblat A.I. USSR #3823818/27-11
Applied on 12-11-84. Published, Bulletin #44-1988. "Discoveries. Inventions ". 1988,#44.

[**13**] Author's Certificate # 1386786 USSR, MKI F 16/15, 8/16. Pipeline Bend. Rozenblat A.I. USSR. #4125878/27-11.Applied on 10-10-84. Published, Bulletin #13-1988. "Discoveries. Inventions ". 1988,#13.

[**14**] Author's Certificate #1164174 USSR, MKI B65G,53/52. Pipeline for pneumatic Transportation of loose materials. Rozenblat A.I. USSR #3501770/27-11. Applied on 10-15-82. Published on 6-30-85, Bulletin #24. "Discoveries. Inventions" 1985. #2, p.84.

[**15**] Author's Certificate #1134504. USSR, MKI B65 G,53/52. Transport Pipeline. Rozenblat A.I.USSR #3451278/27-11. Applied on 6-8-82. Published on 1-15085, Bulletin #2. "Discoveries. Inventions ". 1985,#2, p.84.

[**16**] Anthony V. Nero, Jr. Nuclear Reactors,1979, p.46.—University of California Press.

[**17**] Paul Cohen. Water Coolant Technology of Power Reactor, 1969, p.9.- Gordon and Breach Science Publishers.

[**18**] J.George Wills. Nuclear Power Plant Technology,1967, p.64.- John Willey &Sons, Inc.

[**19**] James F.Connors. Exploring in Aerospace Rocketry,1967, p.62.—National Aeronautics and Space Administration.

[**20**] Winter, Frank H. Rockets Into Space,1990, p.106.- Harvard University Press.

[**21**] Michael Stoiko. Soviet Rocketry: Past, Present, and Future, 1990. p.27. - Holt, Rinehart and Winston.

28 TH ISRAEL CONFERENCE ON MECHANICAL ENGINEERING

1. Statistical methods in evaluation of thermal deformations of work-pieces for the heat-treatment processes.

ABSTRACT

Thermal deformations of thin-walled gear rings for planetary reduction gear are described here.

Oval, the main criterion of thermal deformation, is has been shown to be advantageous in the process of martempering.

187

Diagrams depicting variations in external and internal diameters for the different types of rings are shown before and after martempering.

The results of statistical investigations were made and shown the general linear regressing models.

INTRODUCTION

It is difficult to choose the rational allowance on final metal cutting operations for work-pieces. This is due to absence of reference data for thermal deformations and the character of its distribution in the process of martempering.

These disadvantages considerably increase the labor input of work-pieces, in part because they have high hardness at this period of cutting.

The main purpose of this paper is to show the character of thermal deformations in heat-treatment processes as well as to discover some of its values for the different thin-walled gear rings in the process of martempering.

EXPERIMENTAL STUDIES

The investigations were made on thin-walled gear rings (Figure 1) of planetary reduction gear for marine cranes in accordance with recommendations of author [1] for steel 40X (the same steel AISI 4140).

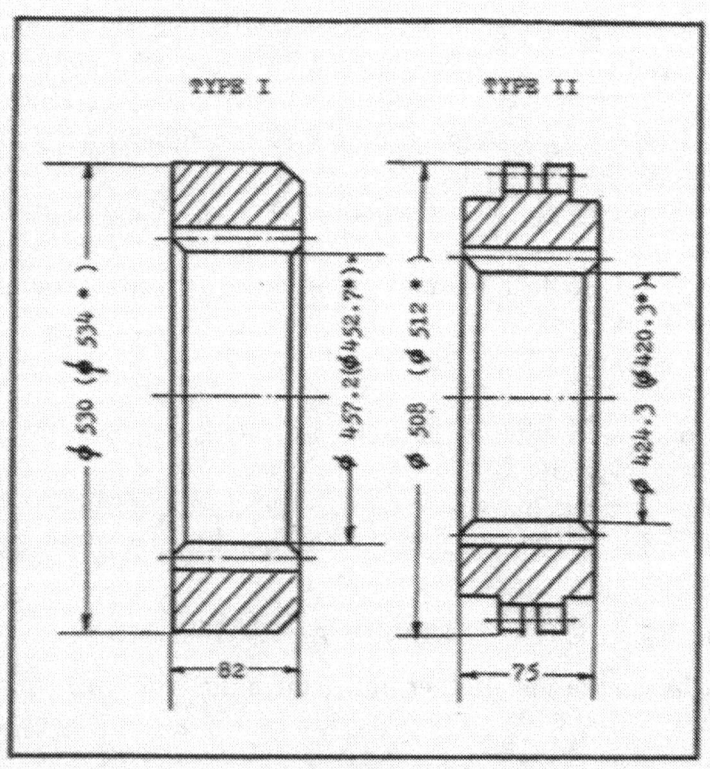

Fig.1 The general types of gear rings for planetary

gear are exposed to heat-treatment (martempering)

(all dimensions in millimetre ; * sizes before heat-treatment)

The main criteria by which the forming detail was evaluated by parameter as ovality of detail. The oval is described as the error of cutting and heat-treatment. This parameter considerably influences on the distribution of allowance in preliminary and final cutting operations.

The following formula for ovality of external and internal diameters was expressed in view of

$$\delta = D_{max} - D_{min} \quad (12)$$

where,

D_{max} = maximum value of diameter;

D_{min} = minimum value of diameter.

Statistical investigations in machine manufacturing showed that the ovality for the cylindrical rings before of heat-treatment, but after the preliminary cutting processes, have the values about of $\delta^0 = 0.05 \div 0.36$ **mm.**

Figure 2 illustrates the diagram of change of the ovality for the external and internal diameters of ring in dependence from type of treatment.

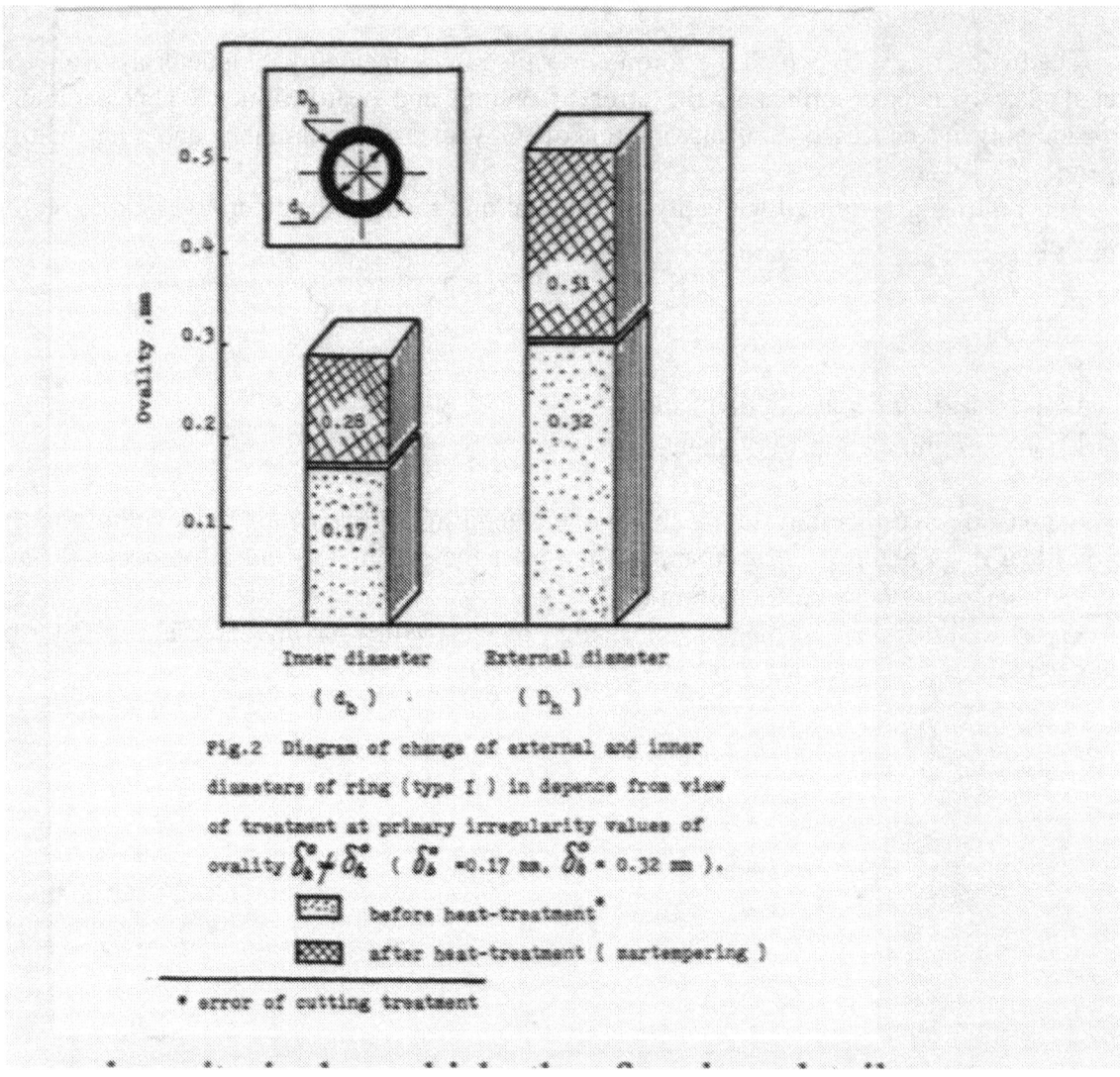

Fig.2 Diagram of change of external and inner
diameters of ring (type I) in depence from view
of treatment at primary irregularity values of
ovality $\delta_a^0 \neq \delta_h^0$ (δ_a^0 =0.17 mm, δ_h^0 = 0.32 mm).

▨ before heat-treatment*

▩ after heat-treatment (martempering)

* error of cutting treatment

From Figure 2 we see that the external and internal diameters have the different values of ovality. Also in Figure 2 is evidence that the primary values of ovality for these diameters increases, particularly after of heat-treatment (martempering and also has an irregular character).

These conclusions are confirmed by the data which are shown in Figure 3.

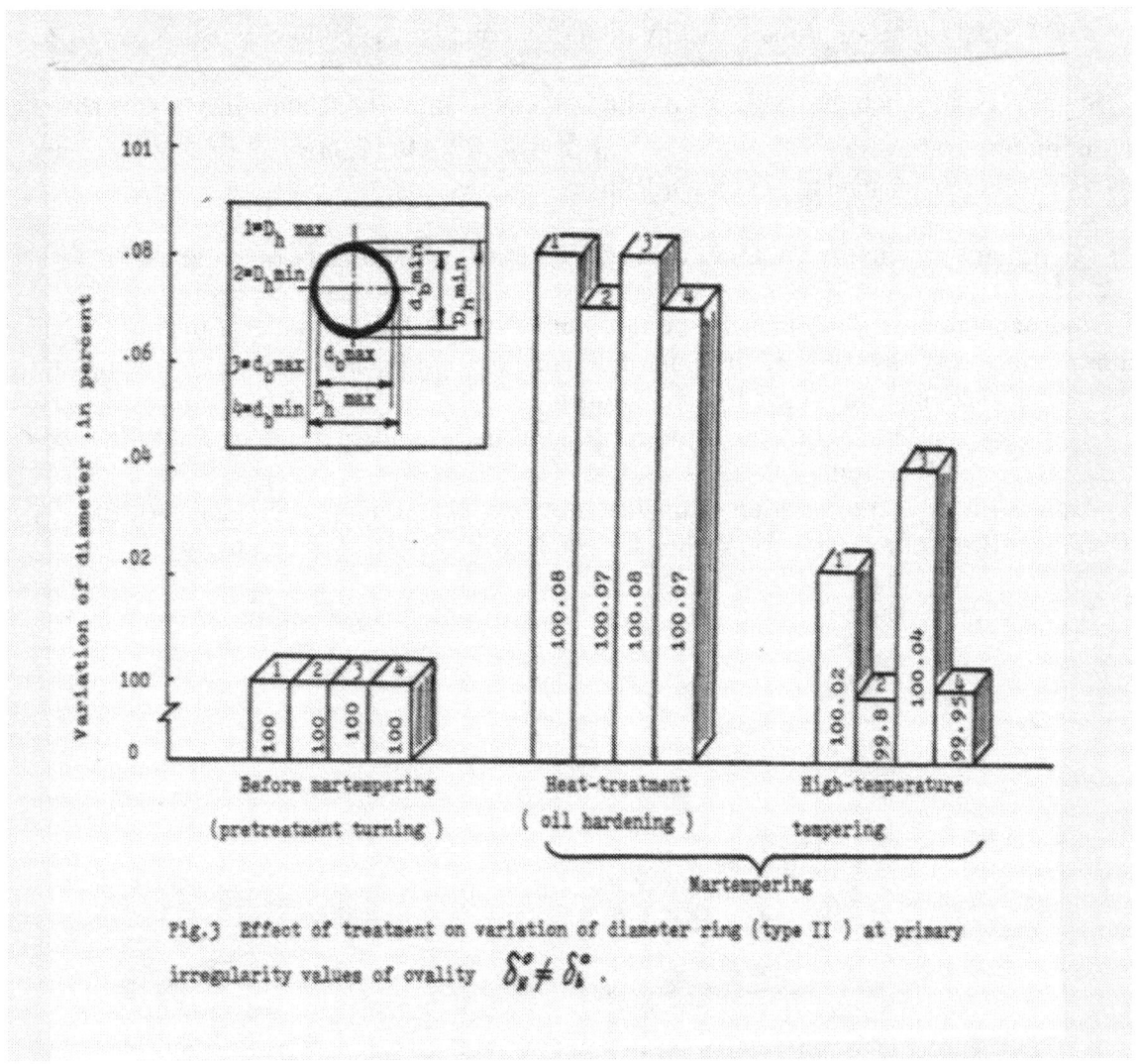

Fig.3 Effect of treatment on variation of diameter ring (type II) at primary irregularity values of ovality $\delta_1^o \neq \delta_2^o$.

From Figure 3 we see that external and internal diameters have an even distribution and insignificance increasing in period of heat-treatment (oil hardening). These changes are joined by phase transformations of steel during of heat-treatment and indicate in fact that the primary value of ovality plays an important role at this moment.

In the process of high-temperature tempering, as shown in Figure 3, there are considerable variations in external and internal diameters. The primary ovality increases on the external and internal diameters into account of the changes in the sizes of the diameters.

The average increment values for diameters in the progress of martempering usual are

equal $\Delta = 0.20 \div 0.80$ mm.

INFLUENCE OF PRIMARY OVALITY ON THE VALUE OF OVALITY AFTER OF MARTEMPERING

Rozenblat, A.I. (1971) in his work **[2]** has showed that the forming of ring after of heat-treatment (martempering) depends on the quality of the ring at the pretreatment turning operation, i.e from the primary ovality of ring and character of its distribution:

a) At equal values of primary ovality on the external $(\delta_H{}^o)$ and internal $(\delta_b{}^o)$

diameters, i.e when the values of ovality are equal $\delta_H{}^o = \delta_b{}^o$. The ovality at this case has the minimum and average its value is equal $\delta' = 0.02 \div 0.10$ **mm,** and has a place in the even distribution, particularly at oil hardening processes.

b) At primary irregularity values of ovality such as when the ovality of external $(\delta_H{}^o)$

and of internal $(\delta_b{}^o)$ diameters are not equal $\delta_H{}^o \neq \delta_b{}^o$. The value of ovality increases considerably, and has the same irregularity character, particularly after of high-temperature tempering (martempering) **[see Figure 2 and 3].**

This appearance may be due to the fact that at high-temperature tempering the water is used as a cooling medium.

Figure 4 depicts the variation of ovality for external and internal diameters of ring (type II) in the process of martempering when the primary ovality has the irregularity

distribution, i.e the condition when the primary ovality has view of $\delta_H{}^o = \delta_b{}^o$.

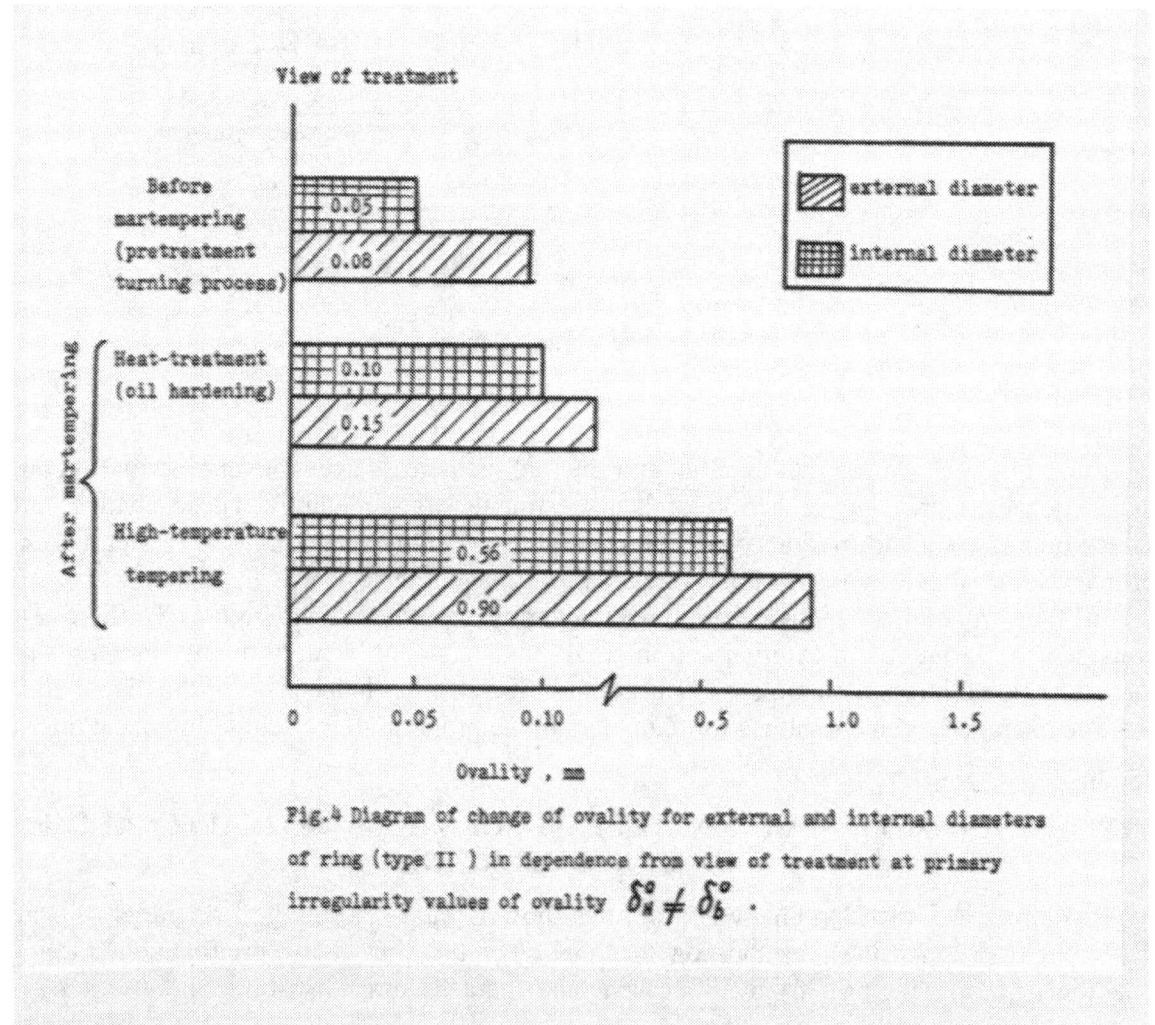

Fig.4 Diagram of change of ovality for external and internal diameters of ring (type II) in dependence from view of treatment at primary irregularity values of ovality $\delta_H^o \neq \delta_b^o$.

Figure 4 illustrates the change of ovality for external and internal diameters of ring (type II). This value is limited to δ'_b =0.10 ÷0.60 mm for internal diameter, and δ'_H =0.15 ÷0.90 mm for external diameters.

Also in Figure 4 is shown the evidence that the value of ovality increases, particularly after of high-temperature tempering when the ring is heated to 520 °C and later have a place the sharp cooling of ring in water as cooling medium.

STATISTICAL ANALYSIS

At primary conditions when the ovality of ring is equal zero ($\delta_H{}^o = \delta_b{}^o$=0), the thermal deformation (ovality) of ring has the uniform distribution. These conclusions are confirmed by data which are shown in Figure 5.

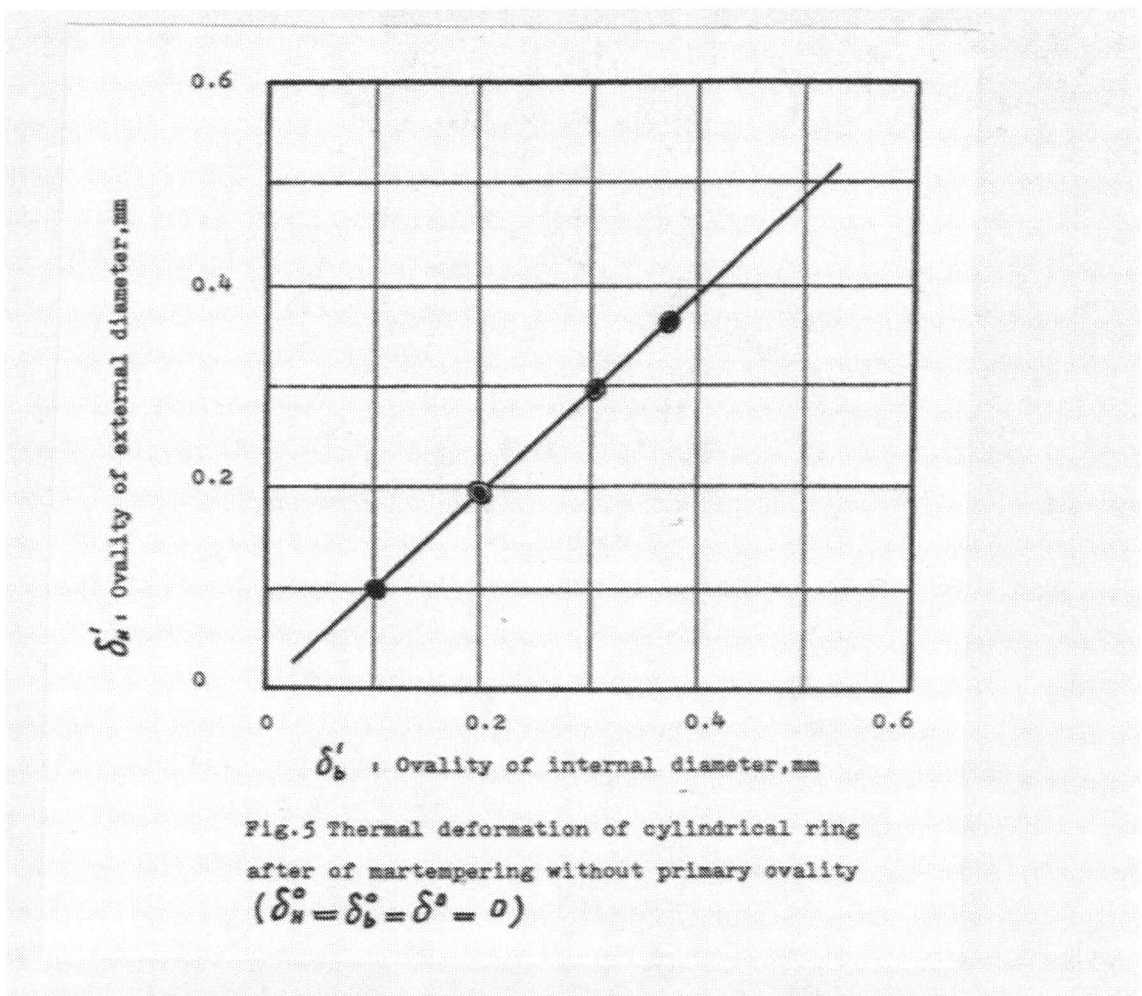

Fig.5 Thermal deformation of cylindrical ring after of martempering without primary ovality $(\delta_H^o = \delta_b^o = \delta^o = 0)$

It is shown in Figure 5 that the ovality of external and internal diameters of ring after of martempering have the same values and is described by linear regression model

$\delta'_H = \delta'_b$. In the presence of uniform primary ovality, i.e at the condition when the values of $\delta_H{}^o = \delta_b{}^o = \delta^o$ and irregularity primary ovality for the external and internal diameters of

ring, i.e at the condition when $\delta_H{}^o \neq \delta_b{}^o$, the ovality of ring increases considerably after of high-temperature tempering (martempering). Figures 6 and 7 illustrate the variation of ovality for the external and internal diameters of ring after martempering.

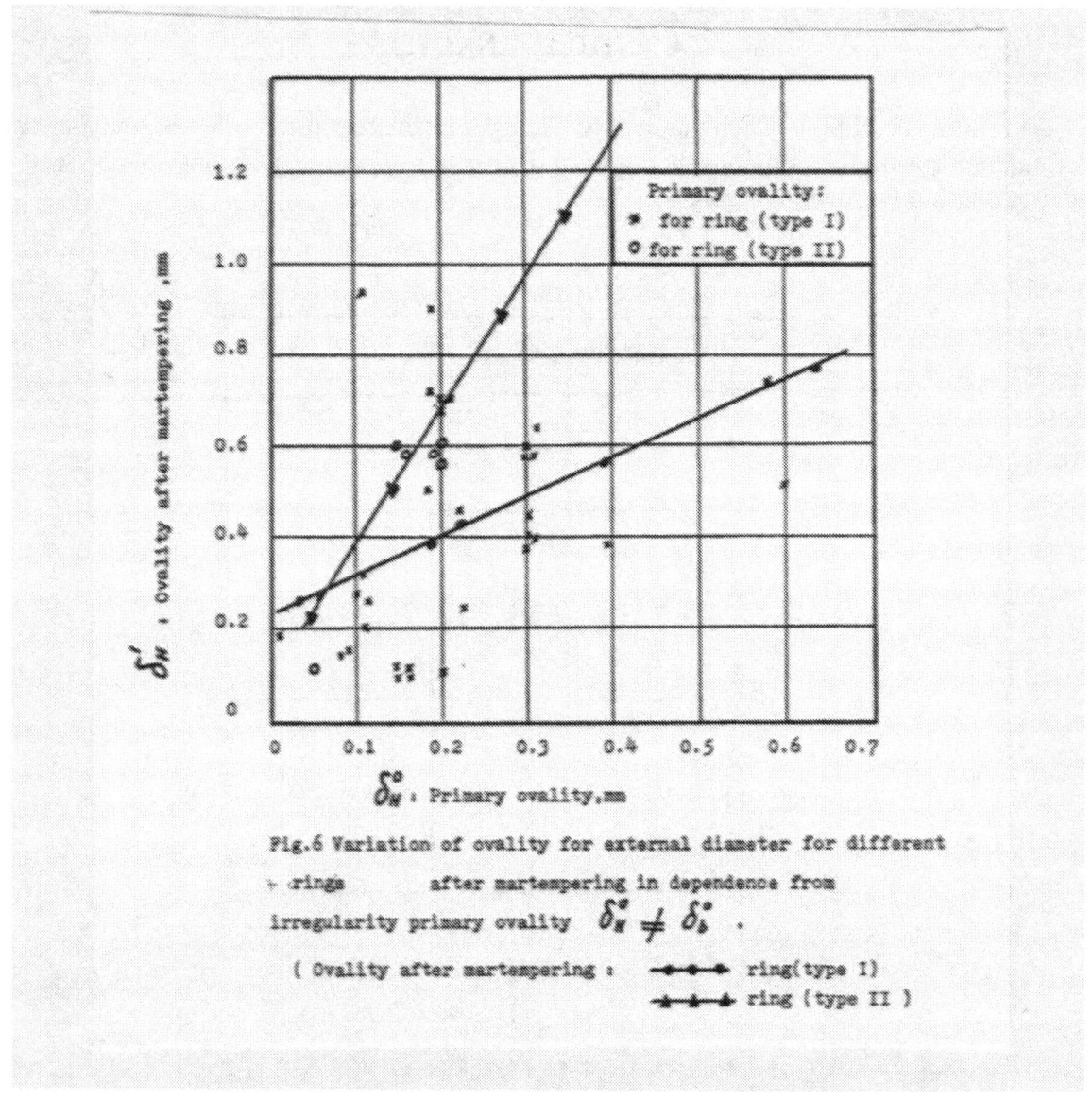

Fig.6 Variation of ovality for external diameter for different rings after martempering in dependence from irregularity primary ovality $\delta_H^o \neq \delta_b^o$.

(Ovality after martempering : ——•—•— ring(type I)
—+—+—+— ring (type II))

Figure 6 shows that variation of ovality for external diameter of ring after of martempering has the linear model. Functional analysis for ovality (δ'_H) of ring (type I) after of martempering in dependence from primary ovality has view of

$$\delta'_H = 0.25 + 0.80\ \delta_H{}^o \quad (2)$$

Analyzing Figure 6, it is shown that the value of ovality for the ring (type I) after of martempering does not exceed value of $\delta'_H = 1.0\ mm$.

Figure 6 also shows that with increasing of the primary ovality, the ovality after of martempering for external diameter of ring increases. This increasing of ovality after of martempering is greater for the ring (type II) than for the ring (type I).

This may be due to the differences on the contour of details. Regression linear model for the ovality (δ'_H) for the ring (type II) after of martempering in dependence from primary ovality has view of

$$\delta'_H = 0.08 + 2.74\ \delta_H^o \ (3)$$

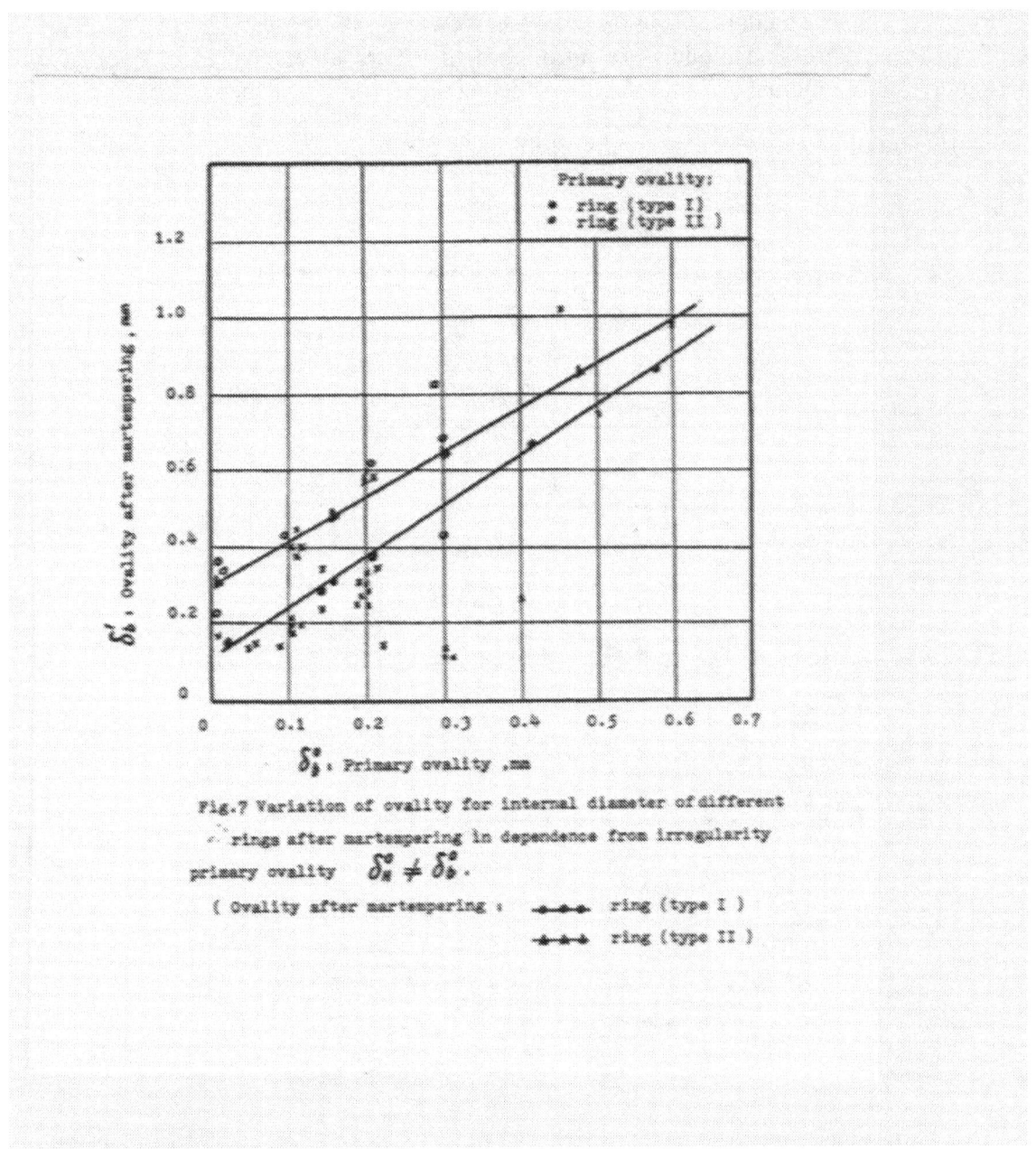

Fig.7 Variation of ovality for internal diameter of different rings after martempering in dependence from irregularity primary ovality $\delta_H^o \neq \delta_b^o$.

(Ovality after martempering : ———— ring (type I)
———— ring (type II)

Figure 7 shows that the variation of ovality for internal diameter of ring after of martempering also has a linear regression model. This functional line given in view of the

formula $\delta'_b = 0.14 + 0.99\ \delta_b\,^o$ (4)

As shown in Figure 7, the scatter plot for the values of ovality after of martempering has the range of **$0.10 \leq \delta'_b \leq 1.0$mm.** The ovality for internal diameter of ring after of martempering increases with increasing of primary ovality.

From Figure 7 the ovality after of martempering is greater for ring (type II) than for ring (type I), as indicated above and this case can be explained due to its complex contour.

Regression linear model for ring (type II) has view $\delta'_b = 0.28 + 1.23\ \delta_b\,^o$ (5)

Two linear regression models are introduced in Figure 8 for change of ovality, after martempering for external and internal diameters of ring (type I) in dependence from

coefficient of primary error at conditions of $\delta_H{}^o \neq \delta_b{}^o$.

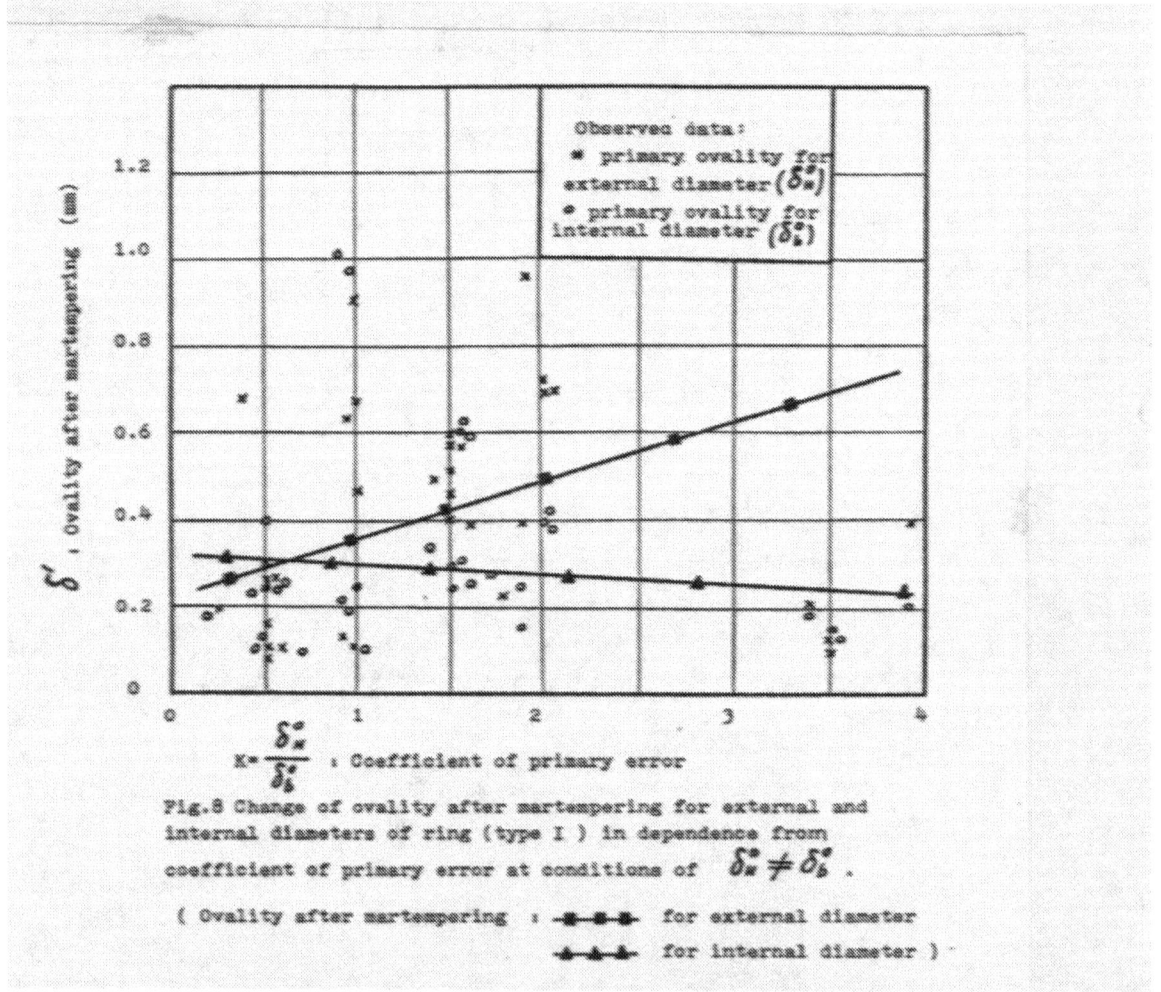

Fig.8 Change of ovality after martempering for external and internal diameters of ring (type I) in dependence from coefficient of primary error at conditions of $\delta_H^o \neq \delta_b^o$.

(Ovality after martempering : ▲▬▲▬▲ for external diameter

▲▬▲▬▲ for internal diameter)

The coefficient at primary error which has place usually after of pretreatment cutting processes is equal to

$$K = \delta_H^o / \delta_b^o \ (6)$$

where,

δ_H^o = primary value of ovality for external diameter;

δ_b^o = primary value of ovality for internal diameter.

Formula (6) indicates that with increasing of value ovality (δ_H^o), the coefficient of

of primary error also increases; with increasing value of ovality (δ_b^o), the coefficient of primary error decreases.

Figure 8 shows that the variation of ovality for external diameter with increasing of coefficient of primary error also increases; variation of ovality for internal diameter with increasing of coefficient of primary error decreases.

These facts indicate that primary ovality plays the most important role on the next deformations of ring in heat-treatment processes. From Figure 8 we see also that linear regression model of ovality for ring (type I) after of martempering for external diameter has view of

$$\delta'_H = 0.142 + 0.01 \ K \ (7)$$

and $\delta'_b = 0.358—0.027 \ K \ (8)$ for internal diameter.

CONCLUSIONS

(1) Thermal deformation for the heat-treatment processes are presented. The multiple regression models are especially useful for these processes at these thermal deformations submit to laws of mathematical statistics.

(2) The results calculated by the present linear regression models used in evaluation of thermal deformations for the thin-walled gear rings show the approximate values which permit one to correctly distribute the total allowance for the future cutting processes.

REFERENCES

[1] Smukov, A.A. Handbook of the heat-treater. Moscow: Mashgiz,1961.

[2] Rozenblat, A.I. "Deformation of cylindrical work-pieces in during of martempering ". Moscow: *Journal of Metallurgy and heat-treatment, vol.8 (1971),68-69.*

2. ANALYSIS OF TWO—STROKE-CYCLE MARINE DIESEL ENGINE IN THE TROPICS

ABSTRACT

A multiple regression model for predicting of the engine speed of two-stroke-cycle marine diesel engine is described here.

The engine speed is evaluated in function of independent variables, such as seawater temperature, duration-in service of ship, wave sea (wind speed and its direction). A total of 183 statistical observation data were made to verify this multiple regression model. Each variable has been analyzed independently under a set of conditions.

It is also shown in this paper that the engine speed of marine diesel changes considerably in the tropics. This paper may also become a guide for designing the new service instructions for people who operate the marine diesel engine in the tropics.

INTRODUCTION

Statistical methods are widely used in the different areas of science and industry, particularly in areas which demands the big investments on realization of experimental and research works. There are many examples in mathematical literature of prosperous use of the methods in biology and life science investigations.

However, there are some scientific works in Marine Engineering which use the mathematical statistics [1] and [2].

The main purpose of this paper to show the peculiarities of operating marine diesel in conditions of sailing ships in the tropical seawaters and wave sea.

On the basis of statistical observation data, using the multiple regression methods, to evaluate the functional dependence of engine speed of two-stroke-cycle diesel from some external factors such as seawater temperature, wave sea (wind speed and its directions) and duration-in service of ship (or its fouling).

BRIEF PECULIARITIES OF OPERATING MARINE ENGINE IN THE TROPICS

The main marine diesel engine in the tropics has some peculiarities:

1. It is known that the exhaust temperature of diesel engine is general criterion of it heat density. As a rule, this parameters increases with the increasing of operation service of diesel engine in the tropical seawaters. The exhaust temperature values sometimes exceed 400 ° C to 430 °C for the two-stroke-cycle marine diesel;
2. In the period of sailing ship in the tropics, the engine speed of diesel decreases, particularly with increasing of operational period. In these conditions, the engine speed periodically changes by operators that to support normal heat density of marine diesel engine. As rule, the high exhaust temperature of diesel engine at this time and necessary to decrease the engine speed in operational process, is the main index that shows that body of ship has the big fouling. The average engine speed usually drops, and its value does not exceed 10 percent for ship sailing in the tropics for more than 6 months;
3. The seawater, using for inner cooling system of diesel engine in the tropics, does not guarantee good operational conditions because the temperature of the seawater is very high, and its value sometimes exceed 25°C to 32°C;
4. The operation of marine diesel engine becomes worse as a result of the presence of air humidity which makes the supercharge engine system unproductive;
5. It is shown by observations that wave sea (wind speed and its directions) worsen the operation of diesel engine in the tropics. It increases the heat density and decreases the service life of engine.

THE GENERAL DIRECTIONS OF SELECTION AND PROCESSING OF STATISTICAL OBSERVATIONAL DATA

Analysis of two-stroke-cycle marine diesel engine in the tropics is based upon the following conditions:

1. All statistical data is observed in operational time of cargo ship, having the main parameters:
   - Deidveit =10,984 ton;
   - Type of main marine diesel engine = two-stroke-cycle (Burmister Wein);

- Horsepower of diesel engine =8,750;
- Average ship speed =14.25 Knots;
- Electro-capacity =3,700 Kw;
- Running area (Black Sea to Indian Ocean and other tropical seas).

2. Statistical observed data of I=183 are treated by the methods of mathematical statistics;
3. In base of independent variables, as the main external factors acting on the operational process of diesel engine, are investigated the following parameters:

X'_1 = seawater temperature, $° C$;

X'_2 = duration-in service of ship (engine), days;

X'_3 = wind speed (Beaufort wind scale);

X'_4 = wind direction, degree.

4. In base of dependent variable is put the engine speed, which is the main index of thermo-dynamical process and heat density of diesel engine, i.e parameter **(Y).**

5. On the basis of functional analysis, having the following characteristics view of

$$Y_1 = \varphi (X_1'); \; Y_2 = \Psi (X'_2); \; Y_3 = \alpha (X'_3); Y_4 = \gamma (X'_4); \text{ and}$$

$$Y = \phi (X'_1; \; X'_2; X'_3; X'_4)$$

Some conclusions and recommendations are made at this paper.

STATISTICAL RESULTS AND DISCUSSION

1. Influence of seawater temperature on the operational process of diesel engine

Figure 1 shows a dependence of engine speed from seawater temperature for the different duration-in service periods of operation diesel engine in the tropics.

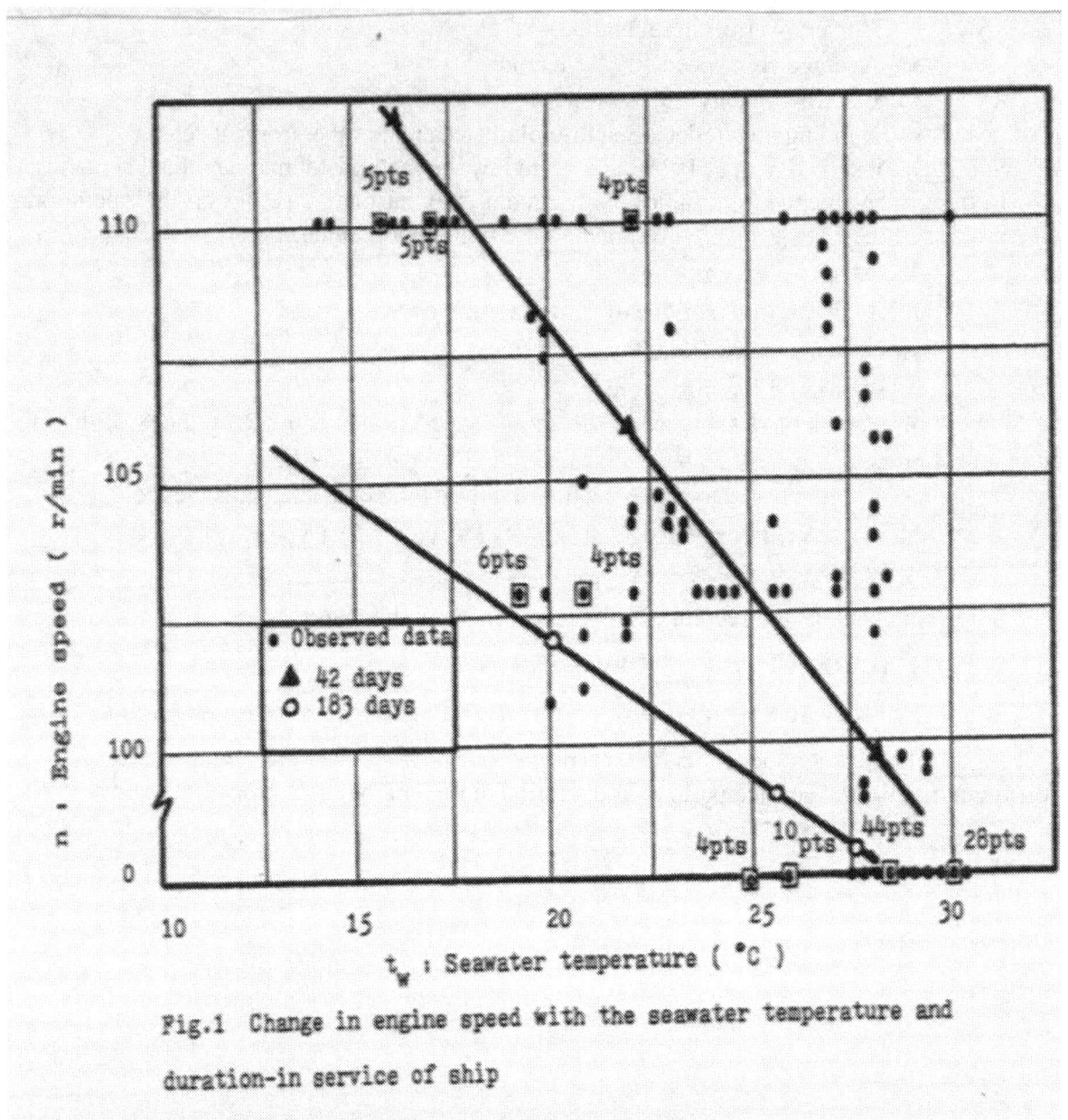

Fig.1 Change in engine speed with the seawater temperature and
duration-in service of ship

It is seen from Figure 1 that in both cases the engine speed decreases with an increasing of seawater temperature and duration-in service. Also from Figure 1, the value of engine speed is smaller for the case when duration-in service is larger.

The residual plot of engine speed against seawater temperature for the different duration-in service of diesel engine is shown in Figure 2. The residual plot from Figure 2 shows that functional dependence view of $Y_1 = \varphi(X'_1)$ follows a linear model.

The analysis of variable ANOVA table which is obtained from different observed data is given in Table 1.

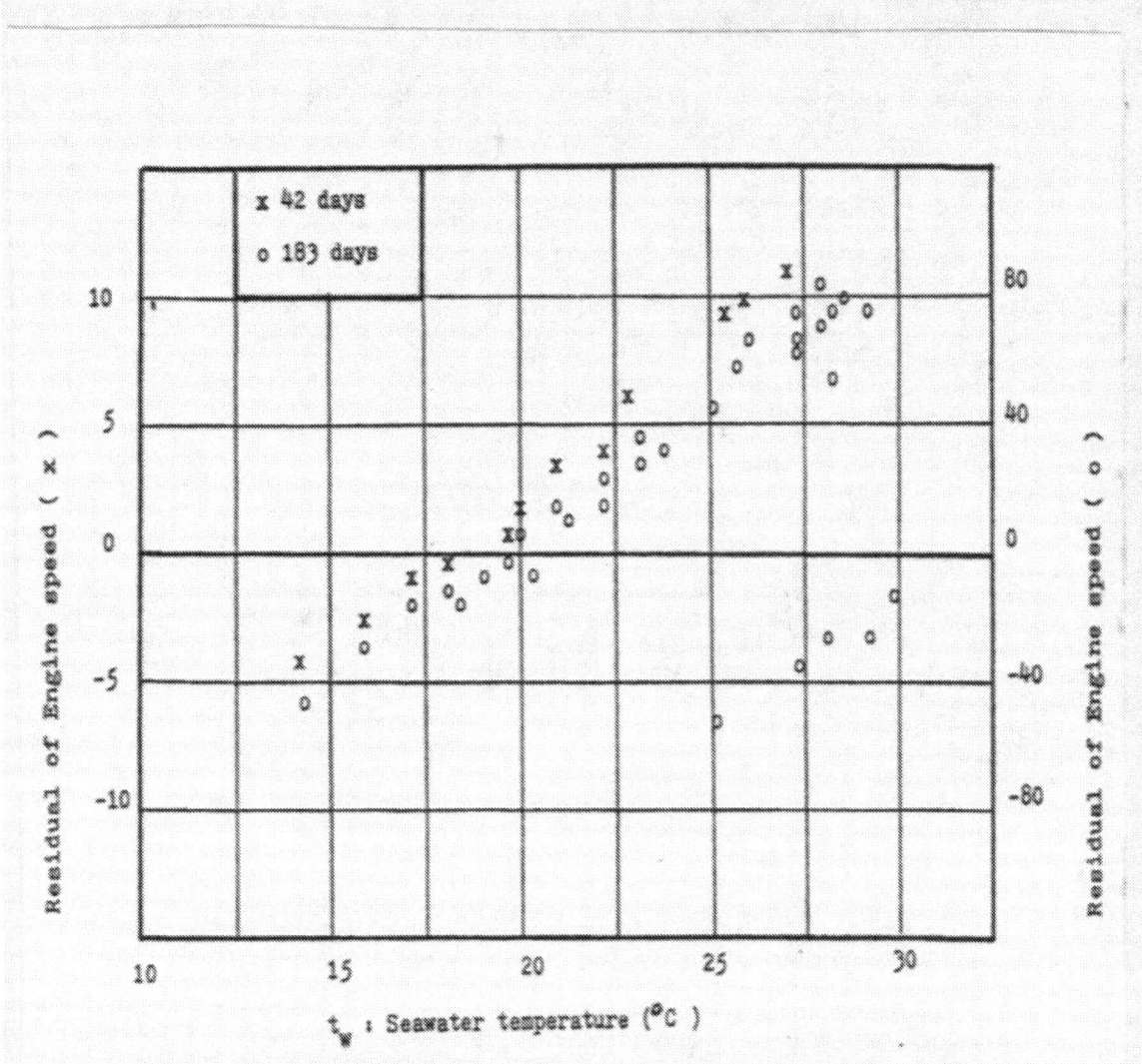

Fig.2 Residual plot of Engine speed against seawater temperature for different days of service

Table 1 ANOVA table for seawater temperature-engine speed different data (i).

Source of variation	$\hat{Y}_1 i = 130.92 - 1.16 X_1 i (i=42)$			$\hat{Y}_1 i = 285.13 - 9.14 X_1 i (i=183)$		
	sum of squares (SS)	degrees of free-dom(df)	mean square (MS)	sum of squares (SS)	degrees of free-dom(df)	mean square (MS)
Regression	1209.16	1	1209.16	231532.3	1	231532.3
Residual	1220.73	40	280.52	273908.10	181	1513.30
Total	12429.89	41		505440.40	182	
Coefficient of correlation	$r = -0.31$			$r' = -0.68$		

It is seen from Table 1 that the coefficient correlation (r = -0.31) is smaller for observed data of i=42 than for the observed data of I=183, as the average seawater temperature in

this case is insignificance $(\overline{X}_{42}) = 21.88$ °C.

It may be concluded that with the increasing of seawater temperature, the fitted regression linear model better describes the relationship between the values of $X_{1,1}$ and

$Y_{1,1}$ when the coefficient of correlation is equal of r= -0.68, the average seawater

temperature is equal of $\overline{X}_{183} = 25.41$ ° C.

2. Connection of engine speed with wave sea (wind speed and its directions).

The influence of wind speed and its direction on engine speed of marine diesel is shown in Figure 3.

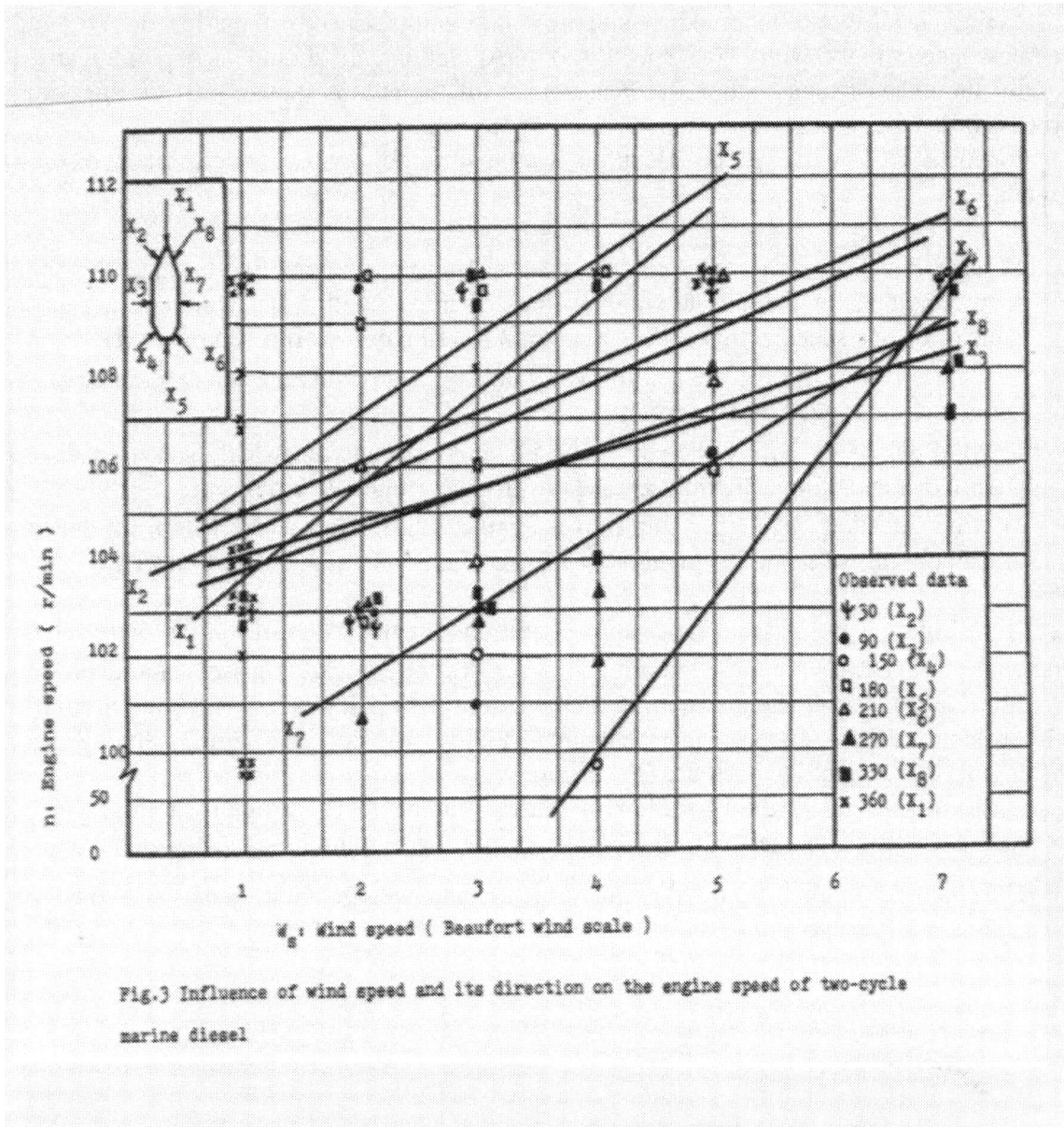

Fig.3 Influence of wind speed and its direction on the engine speed of two-cycle marine diesel

Figure 3 shows that the functional analysis of $Y_3 = \alpha (X'_3)$ and $Y_4 = \gamma(X'_4)$ has linear regression model for any wind directions, acting on the ship in the tropics.

The engine speed in practical conditions demands in the operational processes the decreasing for any wind directions. If the engine speed at this time to accept as the constant or its to increase, as shown in Figure 3, the heat density and exhaust temperature of marine diesel will increase, i.e the conditions for the operation of diesel engine in the tropics and wave sea at this period will worsen.

On the other hand the influence of wind direction on the engine speed is also an important factor.

The observed data from Figure 3 shows that wave sea(wind speed and its directions) has a negative influence on the operation of diesel engine, particularly in conditions when the ship sails in the tropics.

As indicated above, in practical situations the engine speed of marine diesel should decrease accordingly with wind speed and its directions. It should also periodically to control the exhaust temperature and heat density of the engine that to provide the normal process of its operation.

For instance, at wind speed 4 Beaufort, we have the following corrections, as shown in Figure 3:

- At wind direction X_5 primary engine speed should not be corrected because this is the good condition for the marine engine. The ship moves with a tail wind (we accept the value of engine speed at this period about of **n = 110 r/min as the 100 percent);**
- At wind direction X_1, i.e the ship moves against a headwind and this case decrease the primary engine speed by **10 percent;**
- At wind directions X_2, i.e the ship moves against a headwind which acts from the left side. In this situation it should decrease the primary value by **12 percent;**
- At wind direction X_8, i.e the ship moves against a headwind which acts from the right side and in this situation it should decrease the primary it value on **13.3 percent.**

From Figure 3 we see that most unfit conditions for operation of marine diesel engine and ship in the tropics are the wind directions X_2, X_6 and X_4, X_8. They demand decreasing of engine speed, also create the twisting moment, i.e the ship at this moment "turn around".

At this time the operator of marine engine should not to keep the primary engine speed, the engine speed should to decrease (see Figure 3).

These actions make the operation of marine engine more difficult; at this time the exhaust temperature and heat density of diesel engine are very high.

The change in engine speed with wind direction and average wind speed $(\overline{W}_s = 3)$ **is shown in Figure 4.**

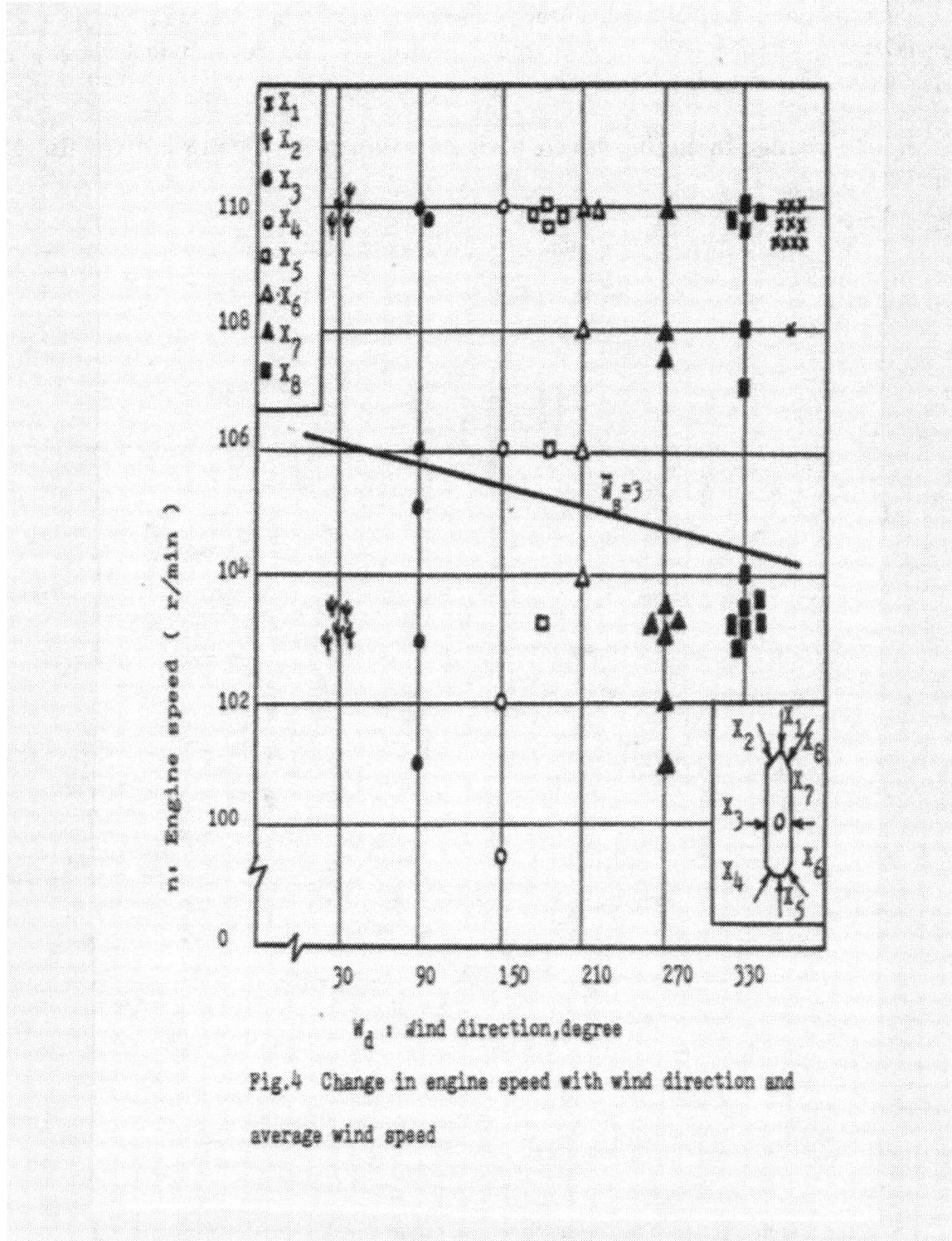

Fig.4 Change in engine speed with wind direction and
average wind speed

This regression function $Y_4 = \gamma (X'_4)$ has the linear model which is described by equation view of

$$X_4 = 106.81 - 0.01 X'_4 \quad (1)$$

Figure 4 also shows the change in engine for the different wind directions. The value of engine speed, as shown in Figure 4, decreases particularly with the wind direction X_1, i.e when the ship moves against a headwind.

3. The change in engine speed with duration-in service of marine diesel in the tropics.

The change of engine speed of marine diesel engine against of duration-in service is shown in Figure 5.

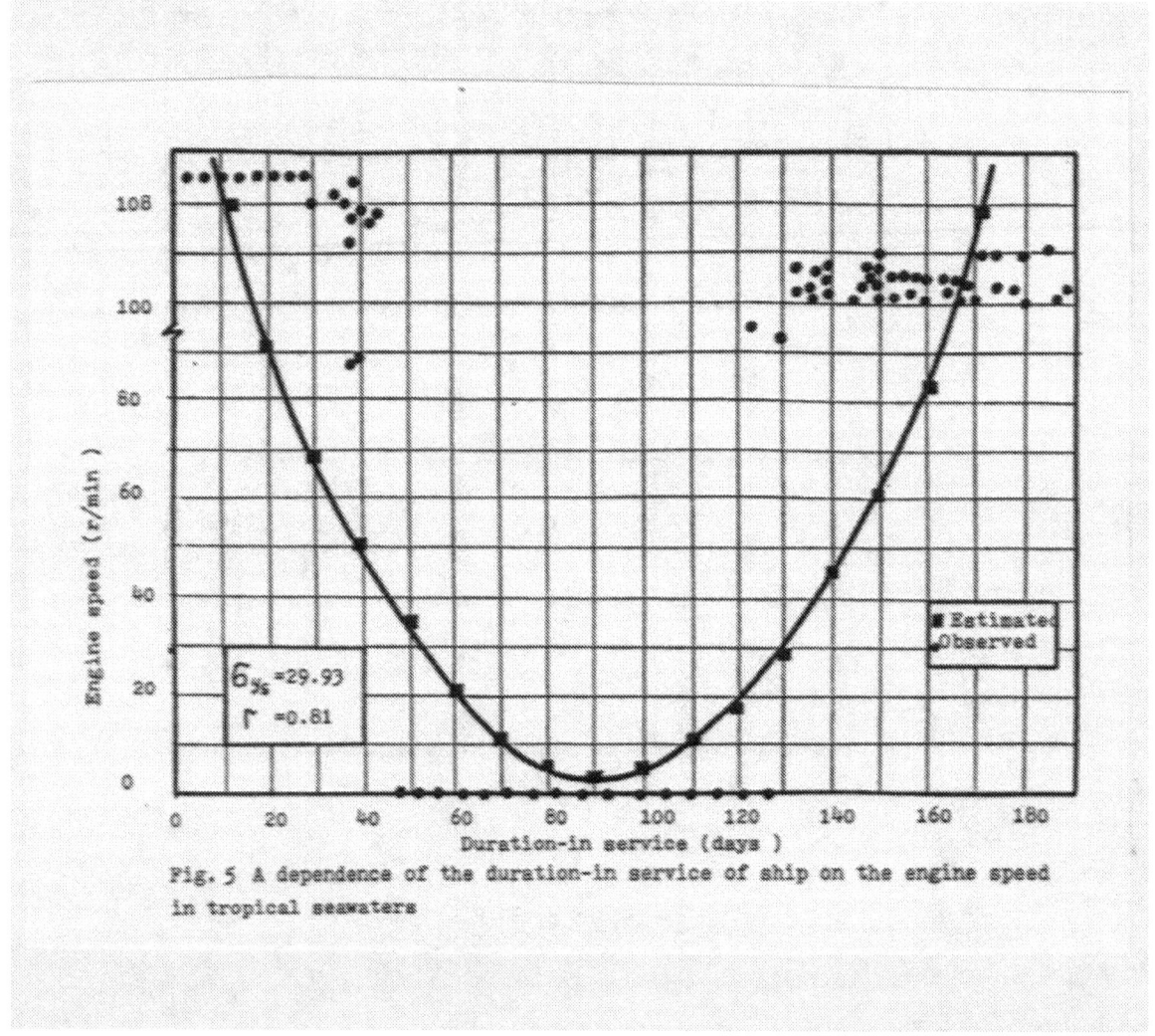

Fig. 5 A dependence of the duration-in service of ship on the engine speed in tropical seawaters

From the Figure 5 it is shown that functional analysis $Y_2 = \Psi (X'_2)$ has the second-order polynomial regression model for one predictor variable X_2 which presents in view of parabola and is described by equation

$$\hat{Y} = 149.27 - 3.13 X^1_2 + 0.02 X^{1\,2}_2 \quad (2)$$

Figure 5 also shows that engine speed decrease with the increasing of duration-in service of ship (diesel engine) in the tropics.

The decreasing of engine speed may be explained by the body ship's fouling allowance. Figure 5 shows a scatter plot (observed data) that distributes three zone:

*The first zone shows the motion of ship in primary conditions (the fouling is equal zero);running area: Black sea to Indian Ocean and other tropical seas with duration—in service of 43 days;

* The second zone the conditions when the ship and marine diesel engine without of operation with duration-in service of 85 days in the tropics;

* The third zone relates to the operational period of marine diesel engine with the duration-in service of 55 days; running area: Indian Ocean and other tropical seas to the Black sea. These zones show that engine speed has the reduction in average of 10 percent relatively of primary conditions.

In Figure 6 is shown the graphical analysis for several different types of residual plots-engine speed versus X_2 for the regression model:

$$\hat{Y} = 149.27 - 3.13\,X'_2 + 0.02\,X^{12}_2 \quad (2\,a)$$

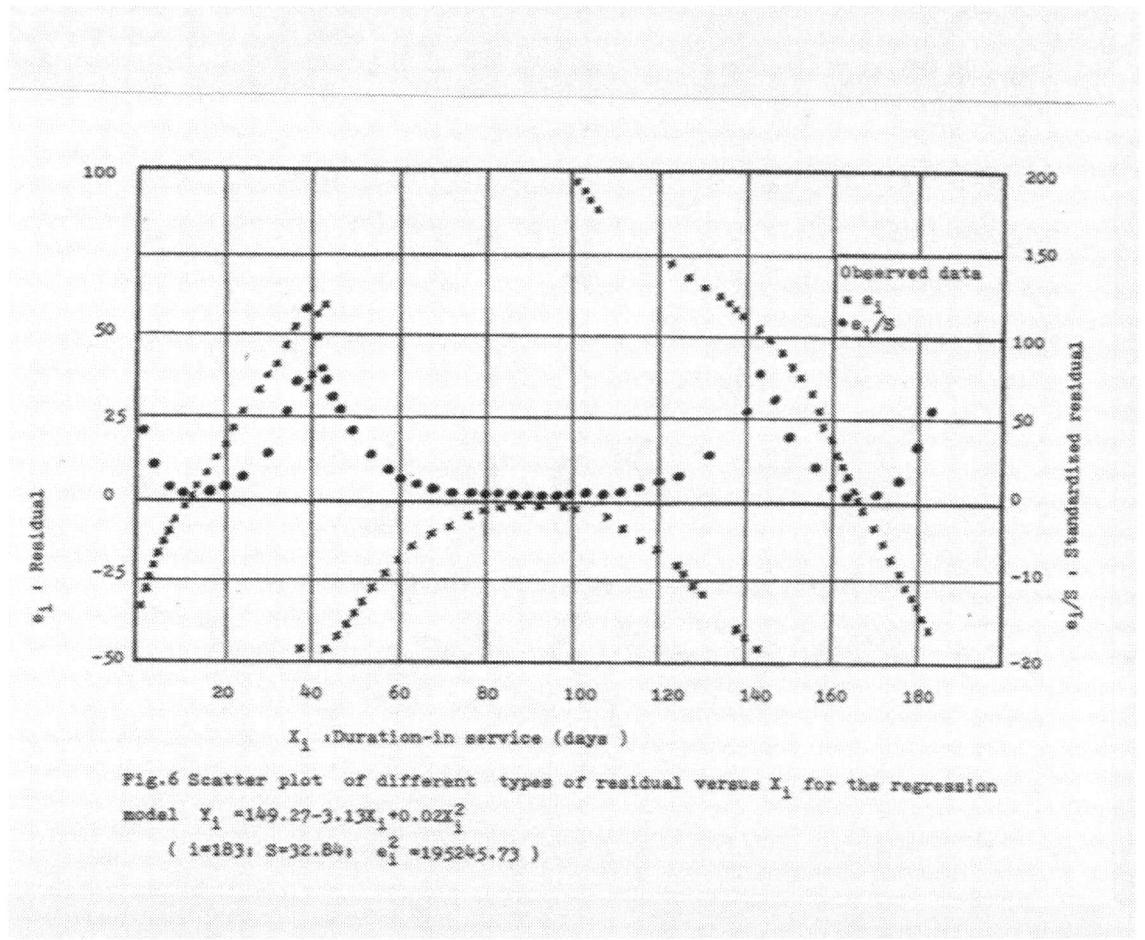

Fig.6 Scatter plot of different types of residual versus X_1 for the regression model $Y_1 = 149.27 - 3.13X_1 + 0.02X_1^2$

(i=183; S=32.84; e_i^2 =195245.73)

4. The multiple regression analysis in evaluation of engine

Functionally, the dependence of engine speed of marine diesel engine in multiple regression analysis has view:

$$Y = \beta_0 + \beta_1 X'_1 + \beta_2 X'_2 + \beta_3 X'_3 + \beta_4 X'_4 \quad (3)$$

Where,

β_0; β_1; β_3 and β_4 are the coefficients of this model.

These coefficients can determine from the system of equations

such as:

$$
\begin{bmatrix}
n & \Sigma X'_{1i} & \Sigma X'_{2i} & \Sigma X'_{3i} & \Sigma X'_{4i} \\
\Sigma X'_{1i} & \Sigma X'^2_{1i} & \Sigma X'_{1i}X'_{2i} & \Sigma X'_{1i}X'_{3i} & \Sigma X'_{1i}X'_{4i} \\
\Sigma X'_{2i} & \Sigma X'_{2i}X'_{1i} & \Sigma X'^2_{2i} & \Sigma X'_{2i}X'_{3i} & \Sigma X'_{2i}X'_{4i} \\
\Sigma X'_{3i} & \Sigma X'_{3i}X'_{1i} & \Sigma X'_{3i}X'_{2i} & \Sigma X'^2_{3i} & \Sigma X'_{3i}X'_{4i} \\
\Sigma X'_{4i} & \Sigma X'_{4i}X'_{1i} & \Sigma X'_{4i}X'_{2i} & \Sigma X'_{4i}X'_{3i} & \Sigma X'^2_{4i}
\end{bmatrix}
\begin{bmatrix}
\beta_0 \\
\beta_1 \\
\beta_2 \\
\beta_3 \\
\beta_4
\end{bmatrix}
=
\begin{bmatrix}
\Sigma Y_i \\
\Sigma X'_{1i}\Sigma Y_i \\
\Sigma X'_{2i}\Sigma Y_i \\
\Sigma X'_{3i}\Sigma Y_i \\
\Sigma X'_{4i}\Sigma Y_i
\end{bmatrix}
$$

$$\Sigma Y_i = n\beta_0 + \beta_1 \Sigma X'_{1i} + \beta_2 \Sigma X'_{2i} + \beta_3 \Sigma X'_{3i} + \beta_4 \Sigma X'_{4i}$$

$$\Sigma X'_{1i}Y_i = \beta_0 \Sigma X'_{1i} + \beta_1 \Sigma X'^2_{1i} + \beta_2 \Sigma X'_{1i}X'_{2i} + \beta_3 \Sigma X'_{1i}X'_{3i} + \beta_4 \Sigma X'_{1i}X'_{4i}$$

$$\Sigma X'_{2i}Y_i = \beta_0 \Sigma X'_{2i} + \beta_1 \Sigma X'_{2i}X'_{1i} + \beta_2 \Sigma X'^2_{2i} + \beta_3 \Sigma X'_{2i}X'_{3i} + \beta_4 \Sigma X'_{2i}X'_{4i}$$

$$\Sigma X'_{3i}Y_i = \beta_0 \Sigma X'_{3i} + \beta_1 \Sigma X'_{3i}X'_{1i} + \beta_2 \Sigma X'_{3i}X'_{2i} + \beta_3 \Sigma X'^2_{3i} + \beta_4 \Sigma X'_{3i}X'_{4i}$$

$$\Sigma X'_{4i}Y_i = \beta_0 \Sigma X'_{4i} + \beta_1 \Sigma X'_{4i}X'_{1i} + \beta_2 \Sigma X'_{4i}X'_{2i} + \beta_3 \Sigma X'_{4i}X'_{3i} + \beta_4 \Sigma X'^2_{4i}$$

As date we have:

$\Sigma Y = 9678$; $n = 183$; $\Sigma X'_1 = 4649.5$;

$\Sigma X'_2 = 16836$; $\Sigma X'_3 = 240$; $\Sigma X'_4 = 24690$;

$\Sigma X'_1 Y = 220561.50$; $\Sigma X'^2_1 = 120901.25$;

$\Sigma X'_1 X'_2 = 431498.50$; $\Sigma X'_1 X'_3 = 5249$;

$\Sigma X'_1 X'_4 = 576585$; $\Sigma X'_2 Y = 902327$;

$\Sigma X'^2_2 = 2059604$; $\Sigma X'_2 X'_3 = 17591$;

$\Sigma X'_2 X'_4 = 2407350$; $\Sigma X'_3 Y = 25545$;

$\Sigma X'^2_3 = 940$; $\Sigma X'_3 X'_4 = 57510$;

$\sum X^1_4 = 2590200$; $\sum X^1_4{}^2 = 7838100$.

The values of coefficients are equal:

$\beta_0 = 76.338$; $\beta_1 = 2.739$; $\beta_2 = 0.076$; $\beta_3 = 9.379$; $\beta_4 = 0.199$.

Thus the multiple regression model express in view of equation:

Yi = 76.338—2.739 X'1, i + 0.076 X'2, i + 9.379 X'3, I + 0.199 X'4, i (4)

In Table 2 is shown an ANOVA table for engine speed (Y_i) regressed on seawater temperature $(X'_{1,I})$, duration-in service $(X'_{2,I})$, wave $(X'_{3,I})$ and wind direction $(X'_{4,I})$.

Table 2 ANOVA table for multiple regression model

$$\hat{Y}_i = 76.338 - 2.739 X'_{1i} + 0.076 X'_{2i} + 9.379 X'_{3i} + 0.199 X'_{4i}$$

Source	Degree of freedom (df)	Sum of squares (SS)	Mean square (MS)	Variance ratio (F)	Coefficient correlation (R)
Regression	4	SSR=445953.76	111488.4	328.61	
Residual	185	SSR=61409.12	339.28		
Total	185	SSY=507362.88			0.938

CONCLUSION

(1) Multiple regression model for evaluation of engine speed of two-strike—cycle marine engine from the different external factors (seawater temperature, duration-in service in the tropics, wave sea(wind speed and its directions) is presented in this paper.

(2) Operation of two-stroke-cycle diesel engine in the tropic showed that the engine speed decreases, and this value will be greater with increasing of duration-in service of ship in the tropical seawaters. For these conditions the average engine speed drops about of 10 percent as the fouling appears on the body of this ship.

(3) And besides the engine speed also decreases as the result of wave sea conditions. At this situation the operator should to reduce the primary engine speed in accordance with the wind speed and its direction that to guarantee the normal heat density of diesel engine.

(4) Regression model in dependence of engine speed against duration-in service has view of polynomial and is described as parabola.

(5) Regression models in dependence of engine speed from wave sea (wind speed and its directions and seawater temperature) have the linear models.

REFERENCES

[1] Perakis, I. "Statistical Analysis of failure Time Distribution for Great Lakes Marine Diesels Using Censored Data". (New York: Society of Naval Architecture and Marine Engineers).*Journal of Ship Research vol.73 (1991):3-5.*

[2] Charnews, L. "Experimental Study of Diesel Engine Cycle-to-Cycle Variation Part I: Analysis of Cycle-to-Cycle cylinder Pressure Variation" (New York: Society of Naval Architecture and Marine Engineers). *Journal of Ship Research, vol.252 (1989):9-12.*

SELECTED BIBLIOGRAPHY

Neter, J. Kutner, M., Nachtsheim, C.Wasserman, W. **Applied Linear Regression Models.**

New York: Richard D.Irving Inc.,1989.

Kleinbaum, D. Kupper, L. Applied Regression Analysis and other multivariable Methods. **New York: Duxbury Press,1978.**

APPENDIX 1

UNION OF SOVIET SOCIALIST REPUBLICS

USSR STATE COMMITTEE ON INVENTIONS AND DISCOVERIES

AUTHOR'S CERTIFICATE

№ 1220856

Authorized by the USSR Government, the USSR State Committee on Inventions and Discoveries issued this Author's Certificate for the invention entitled "**Combination lathe Tool of A. I. Rozenblat**"

Author: **Anatoliy Isaakovich ROZENBLAT**

Applicant: Same as above

Application № 3760953

Priority of invention: July 6, 1984

Registered in State Register of Inventions of USSR on December 1, 1985.

This Author's Certificate is valid throughout the entire territory of the USSR.

S/Chairman of Committee

S/Head of Department

Official Seal

1

PATENT # 1220856 COMBINED LATHE TOOL OF A.I.ROZENBLAT

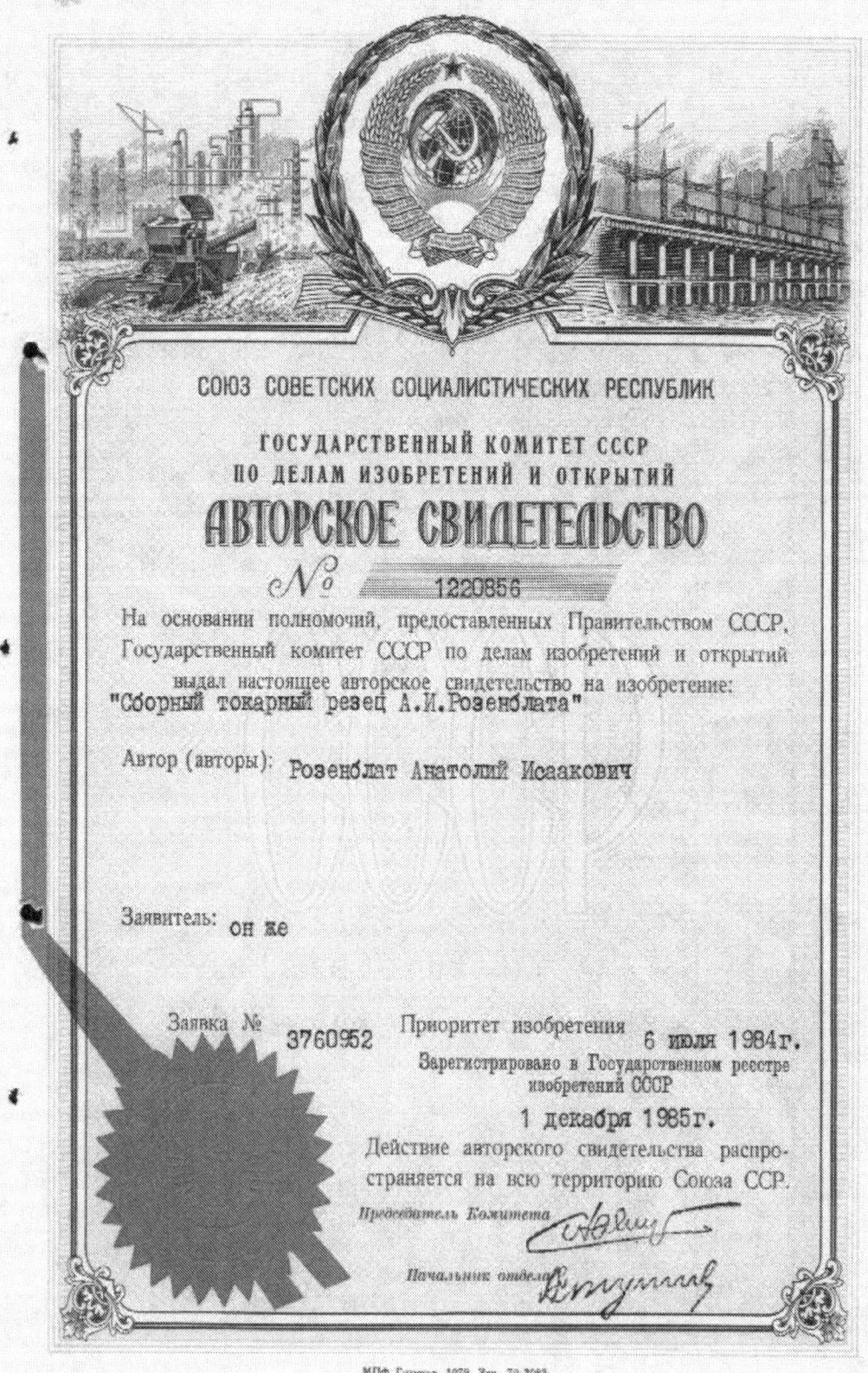

APPENDIX 2

UNION OF SOVIET SOCIALIST REPUBLICS

USSR STATE COMMITTEE ON INVENTIONS AND DISCOVERIES

AUTHOR'S CERTIFICATE

№ 1199466

Authorized by the USSR government, the USSR State committee on Inventions and discoveries issued this Author's Certificate for the invention entitled "Assembling Cutting Tool of A.I. Rozenblat".

Author(s): **Rozenblat** Anatoliy Isaakovich.

Applicant: Same as above

Application № 3766852

Priority of invention: July 6, 1984.

Registered in State Register of Inventions of USSR on August 22, 1985.

This Author's Certificate is valid throughout the entire territory of the USSR.

S/Chairman of Committee

S/Head of Department

Official Seal

Translated from Russian into English

PATENT # 1199466 ASSEMBLING CUTTING TOOL OF A.I.ROZENBLAT

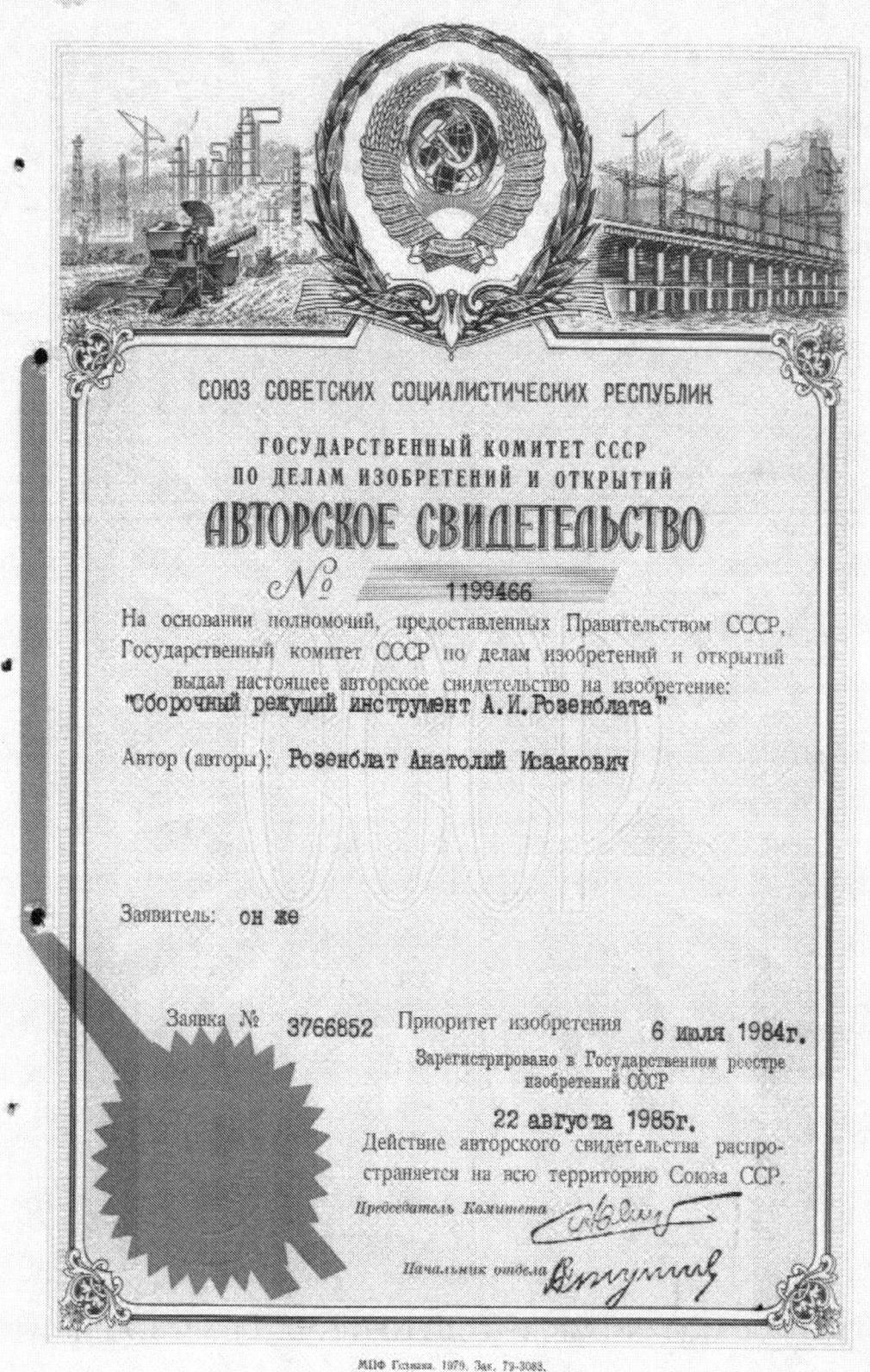

APPENDIX 3

UNION OF SOVIET SOCIALIST REPUBLICS

USSR STATE COMMITTEE ON INVENTIONS AND DISCOVERIES

AUTHOR'S CERTIFICATE

1505676

Authorized by the USSR Government, the USSR State Committee on Inventions and Discoveries issued this Author's Certificate for the invention entitled

"Wrench of A.I.Rozenblat for jaw chuck "

Author :Anatoly Isaakovich Rozenblat

Applicant: Same as above

Application #4291302

Priority of invention:July 28,1987

Registered in State Register of Inventions of USSR on May8,1989.

This Author s Certificate is valid throughout the entire territory of the USSR.

S/Chairman of Committee

S/Head of Department

Official Seal

PATENT # 1505676 WRENCH OF A.I.ROZENBLAT FOR JAW CHUCK

APPENDIX 4

1

UNION OF SOVIET SOCIALIST REPUBLICS

USSR STATE COMMITTEE ON INVENTIONS AND DISCOVERIES

AUTHOR'S CERTIFICATE

№ 1131634

Authorized by the USSR government, the USSR State Committee on Inventions and Discoveries issued this Author's Certificate for the invention entitled "Tool Cleaning Device".

Author(s): **Rozenblat** Anatoliy Isaakovich

Applicant: same as above

Application № 3435998

Priority of invention: May 10, 1982

Registered in State Register of Inventions of USSR on September 1, 1984.

This Author's Certificate is valid throughout the entire territory of the USSR.

S/Chairman of Committee

S/Head of Department

Official Seal

Translated from Russian

PATENT # 1131634 A TOOL CLEANING DEVICE

APPENDIX 5

UNION OF SOVIET SOCIALIST REPUBLICS

USSR STATE COMMITTEE ON INVENTIONS AND DISCOVERIES

AUTHOR'S CERTIFICATE

1504064

Authorized by the USSR Government ,the USSR State Committee on Inventions and Discoveries issued this Author's Certificate for the invention entitled

"Installation of A.I.Rozenblat for cutting of metal bar "

Author:Anatoly Isaakovich Rozenblat

Applicant : Same as above

Application # 4305946

Priority of invention :September 14,1987

Registered in State Register of Inventions of USSR on May 1,1989.

This Author s Certificate is valid throughout the entire territory of the USSR.

S/Chairman of Committee

S/Head of Department

Official Seal

PATENT # 1504064 INSTALLATION OF A.I.ROZENBLAT FOR CUTTING OF METAL BAR

APPENDIX 6

1

UNION OF SOVIET SOCIALIST REPUBLICS

USSR STATE COMMITTEE ON INVENTIONS AND DISCOVERIES

AUTHOR'S CERTIFICATE

№ 1386786

Authorized by the USSR government, the USSR State committee on Inventions and discoveries issued this Author's Certificate for the invention entitled "Pipeline Bend".

Author(s):   **Rozenblat** Anatoliy Isaakovich, **Kopanev** Dmitriy Borisovich and **Vasnev** Anatoliy Ivanovich

Applicant: SCIENTIFIC PRODUCTION ASSOCIATION SPETSTEKHOSNASTKA

Application № 4125873

Priority of invention: October 10, 1986.

Registered in State Register of Inventions of USSR on December 8, 1987

This Author's Certificate is valid throughout the entire territory of the USSR.

S/Chairman of Committee

S/Head of Department

Official Seal

Translated from Russian into English

PATENT # 1386786 PIPELINE BEND

221

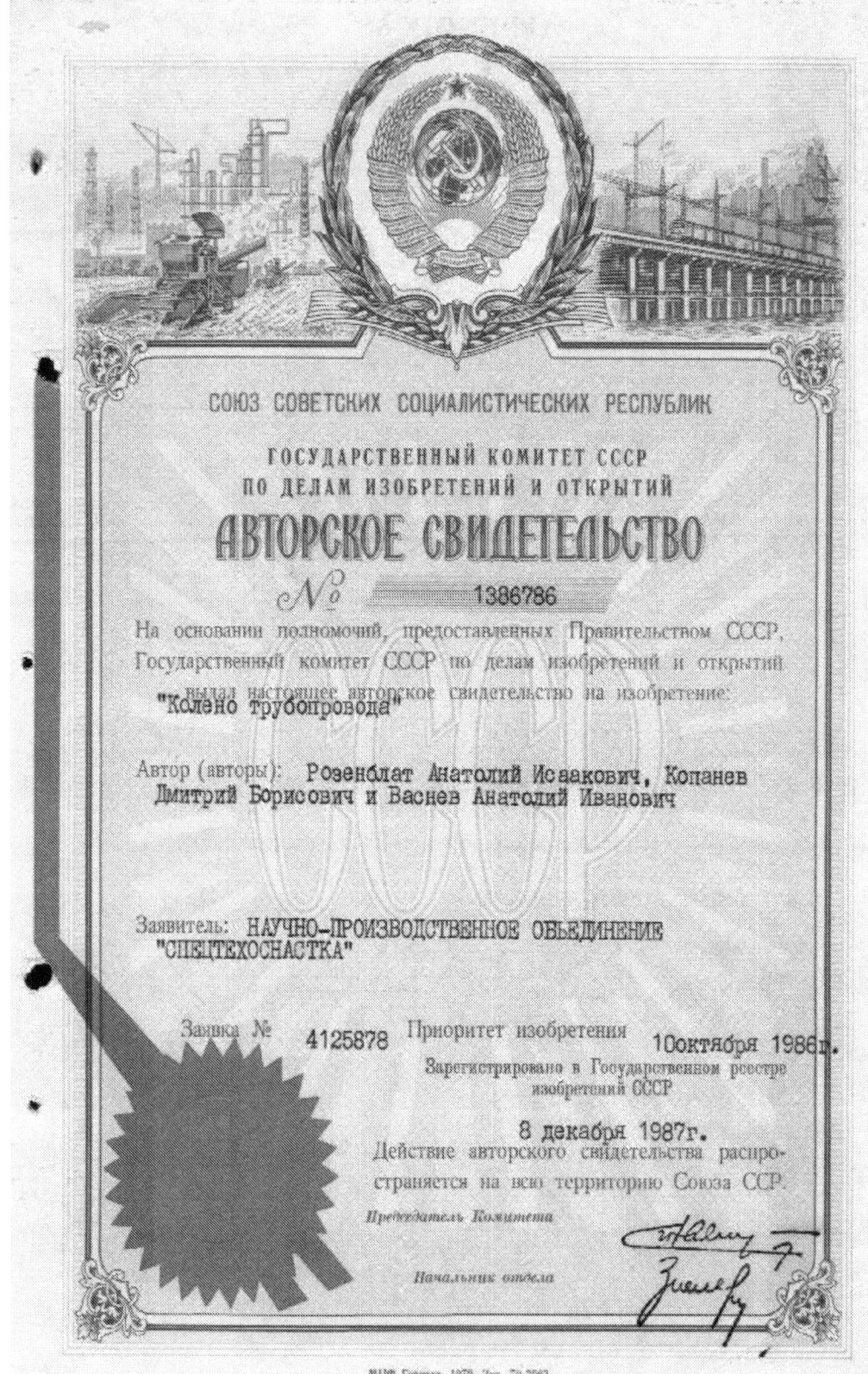

СОЮЗ СОВЕТСКИХ СОЦИАЛИСТИЧЕСКИХ РЕСПУБЛИК

ГОСУДАРСТВЕННЫЙ КОМИТЕТ СССР
ПО ДЕЛАМ ИЗОБРЕТЕНИЙ И ОТКРЫТИЙ

АВТОРСКОЕ СВИДЕТЕЛЬСТВО

№ 1386786

На основании полномочий, предоставленных Правительством СССР, Государственный комитет СССР по делам изобретений и открытий выдал настоящее авторское свидетельство на изобретение:

"Колено трубопровода"

Автор (авторы): Розенблат Анатолий Исаакович, Копанев Дмитрий Борисович и Васнев Анатолий Иванович

Заявитель: НАУЧНО-ПРОИЗВОДСТВЕННОЕ ОБЪЕДИНЕНИЕ "СПЕЦТЕХОСНАСТКА"

Заявка № 4125878   Приоритет изобретения

10 октября 1986 г.

Зарегистрировано в Государственном реестре изобретений СССР

8 декабря 1987 г.

Действие авторского свидетельства распространяется на всю территорию Союза ССР.

Председатель Комитета

Начальник отдела

МПФ Гознака. 1979. Зак. 79-3083.

APPENDIX 7

Union of Soviet Socialist Republics

USSR STATE COMMITTEE ON INVENTIONS AND DISCOVERIES

AUTHOR'S CERTIFICATE

360155

Authorized by the USSR Government ,the USSR State Committee on Inventions
and Discoveries issued this Author's Certificate for invention entitled

LATHE TOOL

Author : Anatoly Isaakovich Rozenblat

Applicant : Same as above

Application # 1364370

Priority of invention : August 15,1969

Registered in State Register of Inventions of USSR on September 5,1972.
This Author's Certificate is valid throughout the entire territory of
the USSR.

S/Chairman of Committee
S/Head of Department
Official Seal

LATHE TOOL

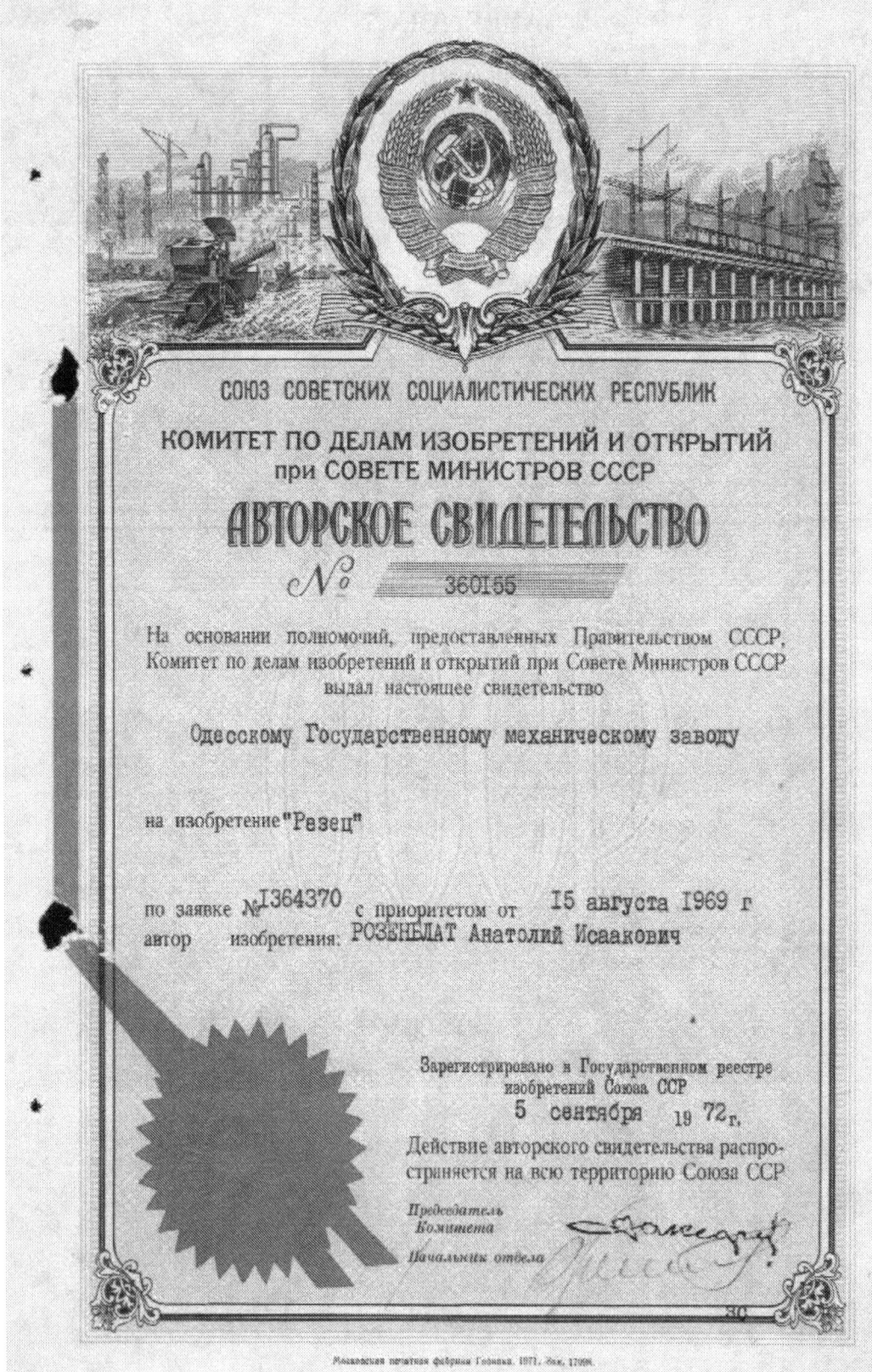

СОЮЗ СОВЕТСКИХ СОЦИАЛИСТИЧЕСКИХ РЕСПУБЛИК

КОМИТЕТ ПО ДЕЛАМ ИЗОБРЕТЕНИЙ И ОТКРЫТИЙ
при СОВЕТЕ МИНИСТРОВ СССР

АВТОРСКОЕ СВИДЕТЕЛЬСТВО

№ 360155

На основании полномочий, предоставленных Правительством СССР,
Комитет по делам изобретений и открытий при Совете Министров СССР
выдал настоящее свидетельство

Одесскому Государственному механическому заводу

на изобретение "Резец"

по заявке № 1364370 с приоритетом от 15 августа 1969 г
автор изобретения: РОЗЕНБЛАТ Анатолий Исаакович

Зарегистрировано в Государственном реестре
изобретений Союза ССР

5 сентября 19 72 г.

Действие авторского свидетельства распространяется на всю территорию Союза ССР

Председатель
Комитета

Начальник отдела

APPENDIX 8

UNION OF SOVIET SOCIALIST REPUBLICS
USSR STATE COMMITTEE ON INVENTIONS AND DISCOVERIES

AUTHOR'S CERTIFICATE

164174

Authorized by the USSR Government,the USSR State Committee on Inventions
and Discoveries issued this Author's Certificate for the invention
entitled
      "Pipeline for pneumatic transportation of granular materials "

Author: Anatoly Isaakovich Rozenblat
Applicant: Same as above

Application #3501770 October 15,1987

Priority of invention:
Registered in State Register of Inventions of USSR on March 1,1985.
This Author's Certificate is valid throughout the entire territory of
the USSR.

      S/Chairman of Committee
      S/Head of Department
      Official Seal

**PATENT # 1164174 PIPELINE FOR PNEUMATIC TRANSPORTATION OF
GRANULAR MATERIALS**

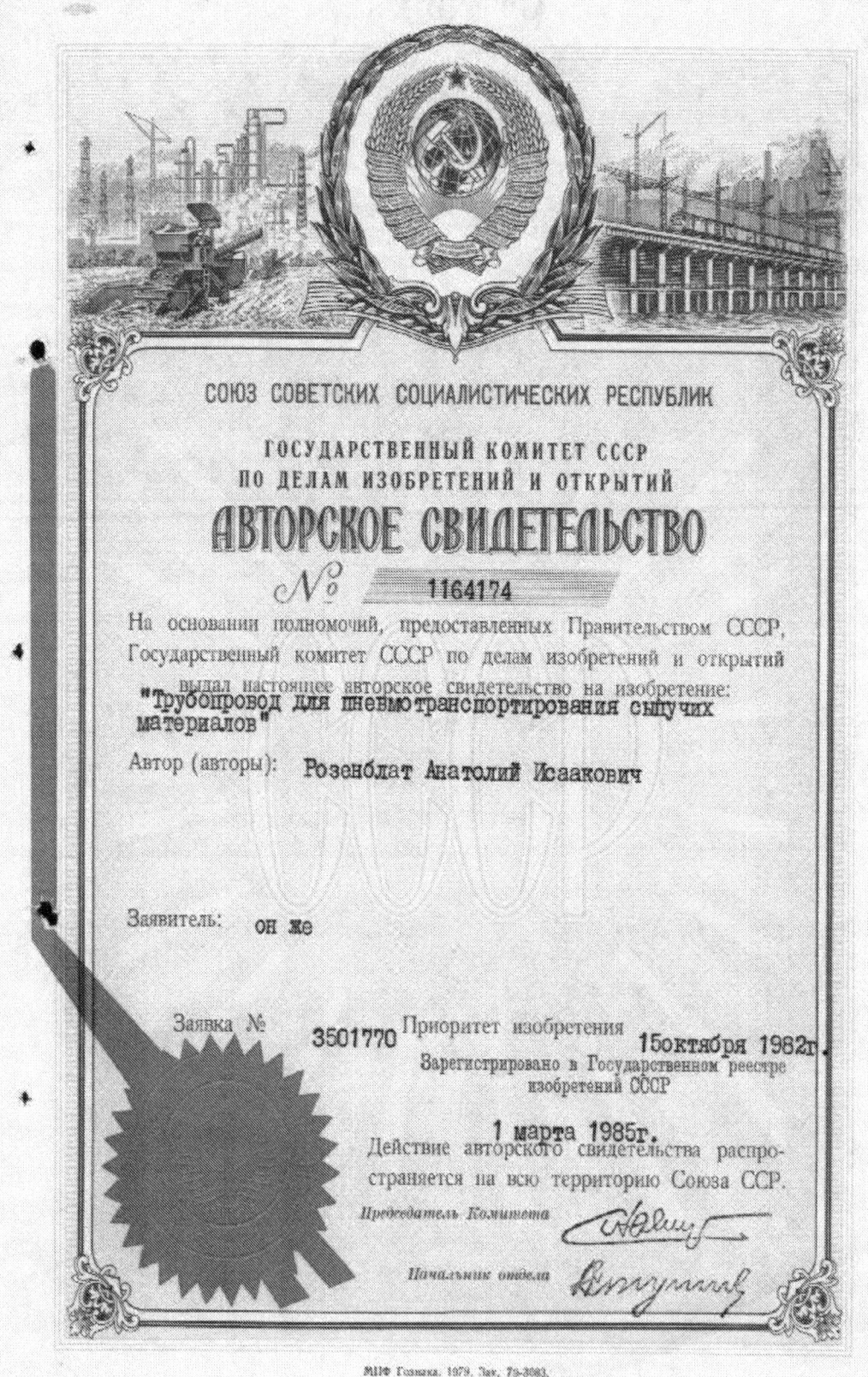

СОЮЗ СОВЕТСКИХ СОЦИАЛИСТИЧЕСКИХ РЕСПУБЛИК

ГОСУДАРСТВЕННЫЙ КОМИТЕТ СССР
ПО ДЕЛАМ ИЗОБРЕТЕНИЙ И ОТКРЫТИЙ

АВТОРСКОЕ СВИДЕТЕЛЬСТВО

№ 1164174

На основании полномочий, предоставленных Правительством СССР, Государственный комитет СССР по делам изобретений и открытий выдал настоящее авторское свидетельство на изобретение:
"Трубопровод для пневмотранспортирования сыпучих материалов"

Автор (авторы): **Розенблат Анатолий Исаакович**

Заявитель: он же

Заявка № 3501770 Приоритет изобретения 15 октября 1982 г.
Зарегистрировано в Государственном реестре изобретений СССР

1 марта 1985 г.
Действие авторского свидетельства распространяется на всю территорию Союза ССР.

Председатель Комитета

Начальник отдела

МПФ Гознака. 1979. Зак. 79-3083.

APPENDIX 9

UNION OF SOVIET SOCIALIST REPUBLICS

USSR STATE COMMITTEE ON INVENTIONS AND DISCOVERIES

AUTHOR'S CERTIFICATE

№ 1441658

Authorized by the USSR Government, the USSR State Committee on Inventions and Discoveries issued this Author's Certificate for the invention entitled "**Method of Processing Curvilinear Channels of A.I. Rozenblat**".

Author: **Anatoliy Isaakovich ROZENBLAT**

Applicant: Same as above

Application № 3823818

Priority of invention: December 11, 1984

Registered in State Register of Inventions of USSR on August 1, 1988

This Author's Certificate is valid throughout the entire territory of the USSR.

S/Chairman of Committee

S/Head of Department

Official Seal

1

PATENT # 1441658 METHOD OF PROCESSING CURVILINEAR CHANNELS OF A.I.ROZENBLAT

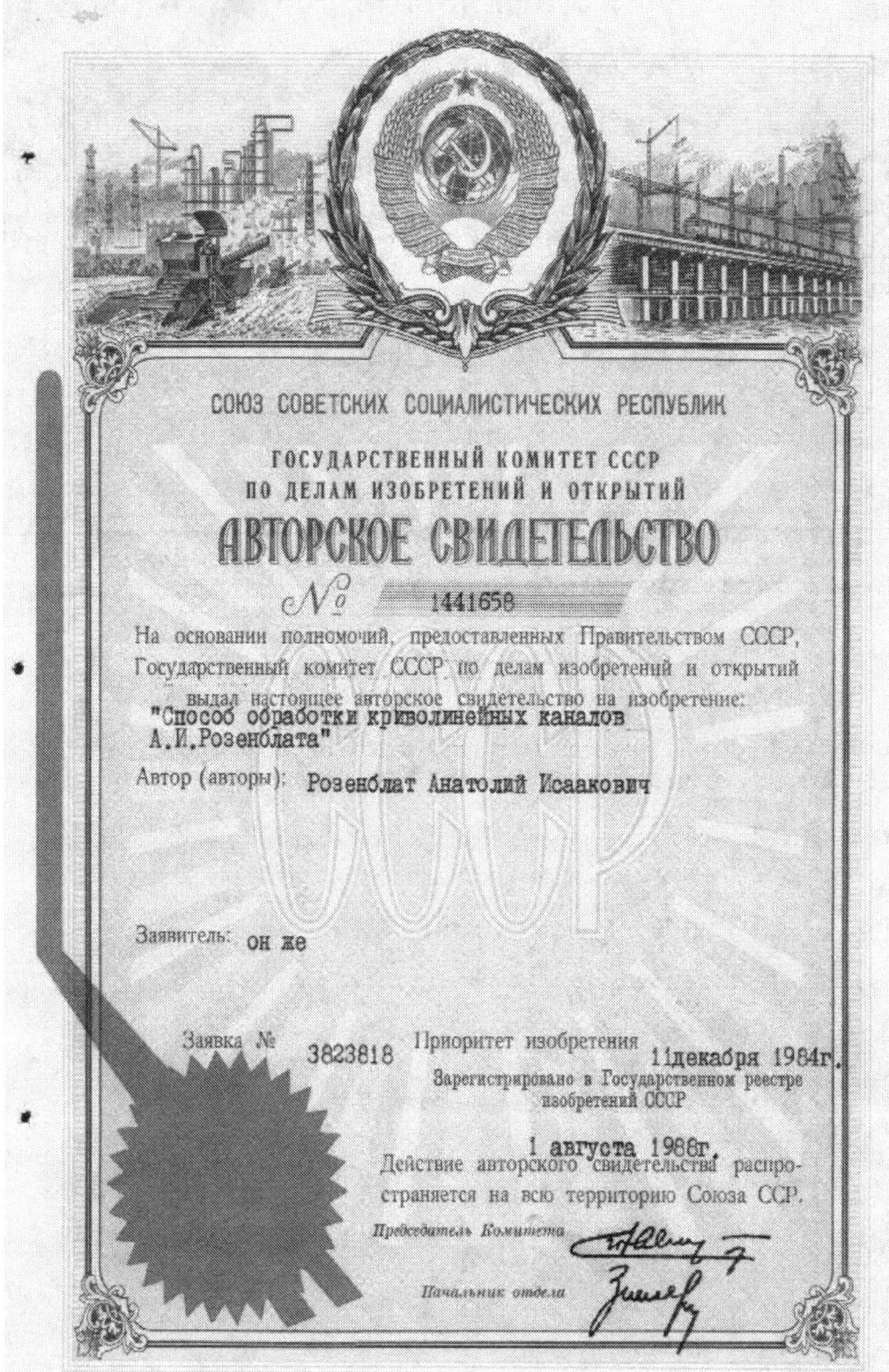

APPENDIX 10

UNION OF SOVIET SOCIALIST REPUBLICS

USSR STATE COMMITTEE ON INVENTIONS AND DISCOVERIES

AUTHOR'S CERTIFICATE

1500416

Authorized by the USSR Government ,the USSR State Committee on Inventions and Discoveries issued this Author's Certificate for the invention entitled

"Multi-positional combined die of A.I.Rozenblat for cutting of sheet materials "

Author : Anatoly Isaakovich Rozenblat

Applicant : Same as above

Application # 4354571

Priority of invention : December 8,1987

Registered in State Register of Inventions of USSR on April 15,1989.

This Author's Certificate is valid throughout the entire territory of the USSR.

S/Chairman of Committee

S/Head of Department

Official Seal

PATENT #1500416 MULTI-POSITIONAL COMBINED DIE OF A.I.ROZENBLAT FOR CUTTING OF SHEET MATERIALS

СОЮЗ СОВЕТСКИХ СОЦИАЛИСТИЧЕСКИХ РЕСПУБЛИК

ГОСУДАРСТВЕННЫЙ КОМИТЕТ СССР
ПО ДЕЛАМ ИЗОБРЕТЕНИЙ И ОТКРЫТИЙ

АВТОРСКОЕ СВИДЕТЕЛЬСТВО

№ 1500416

На основании полномочий, предоставленных Правительством СССР,
Государственный комитет СССР по делам изобретений и открытий
выдал настоящее авторское свидетельство на изобретение:
"Многопозиционный комбинированный штамп А.И.
Розенблата для обработки листовых материалов"

Автор (авторы): **Розенблат Анатолий Исаакович**

Заявитель: он же

Заявка № 4354571   Приоритет изобретения

8 декабря 1987 г.

Зарегистрировано в Государственном реестре
изобретений СССР

15 апреля 1989 г.

Действие авторского свидетельства распространяется на всю территорию Союза ССР.

Председатель Комитета

Начальник отдела

МПФ Госзнака, 1979. Зак. 79-30363.

230

APPENDIX 11

1

UNION OF SOVIET SOCIALIST REPUBLICS

USSR STATE COMMITTEE ON INVENTIONS AND DISCOVERIES

AUTHOR'S CERTIFICATE

№ 1134504

Authorized by the USSR government, the USSR State Committee on Inventions and Discoveries issued this Author's Certificate for the invention entitled "Transport Pipeline".

Author(s): **Rozenblat** Anatoliy Isaakovich

Applicant: same as above

Application № 3451278

Priority of invention: June 8, 1982.

Registered in State Register of Inventions of USSR on September 15, 1984.

This Author's Certificate is valid throughout the entire territory of the USSR.

S/Chairman of Committee

S/Head of Department

Official Seal

Translated from Russian

PATENT # 1134504 TRANSPORT PIPELINE

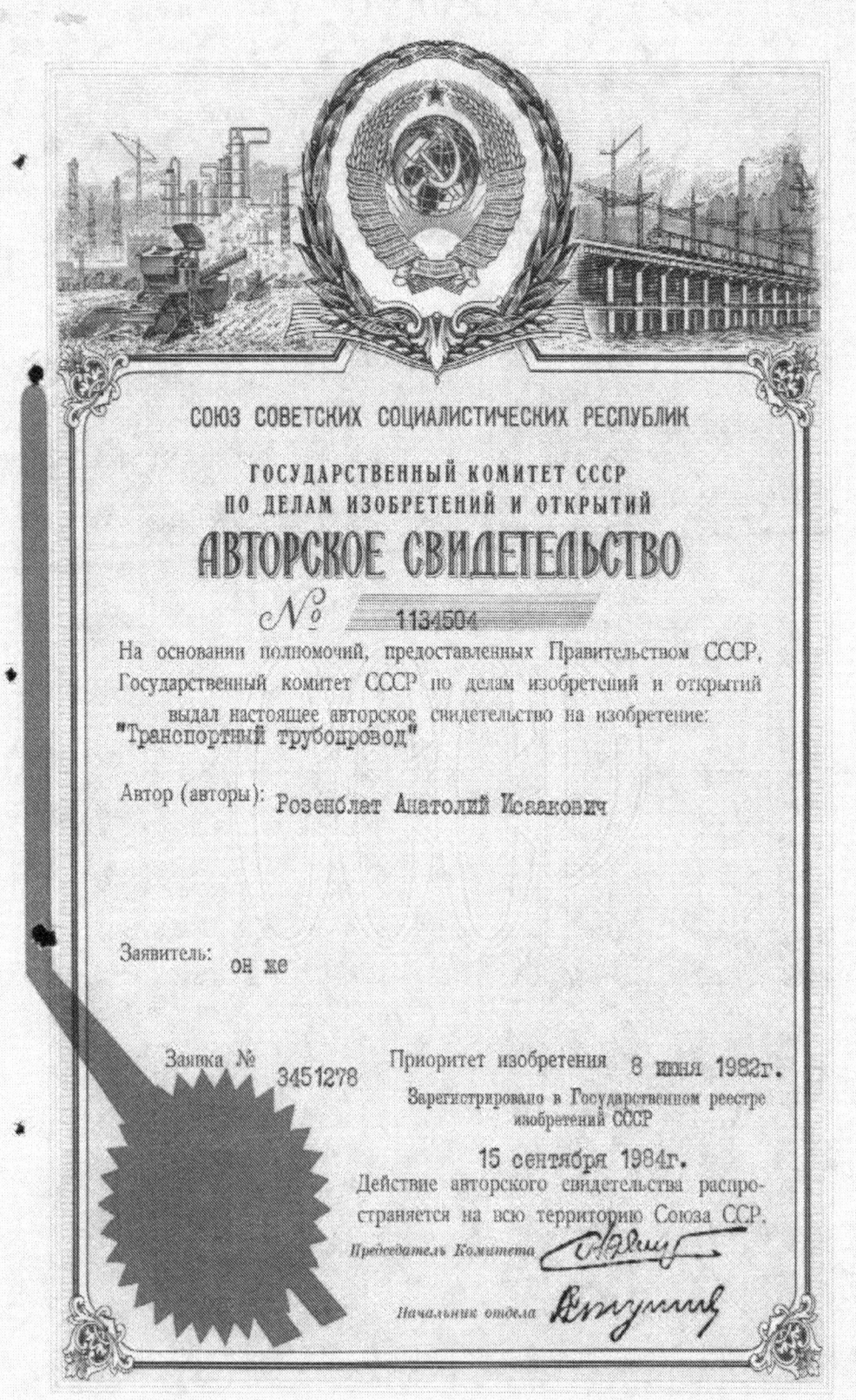

СОЮЗ СОВЕТСКИХ СОЦИАЛИСТИЧЕСКИХ РЕСПУБЛИК

ГОСУДАРСТВЕННЫЙ КОМИТЕТ СССР
ПО ДЕЛАМ ИЗОБРЕТЕНИЙ И ОТКРЫТИЙ

АВТОРСКОЕ СВИДЕТЕЛЬСТВО

№ 1134504

На основании полномочий, предоставленных Правительством СССР,
Государственный комитет СССР по делам изобретений и открытий
выдал настоящее авторское свидетельство на изобретение:
"Транспортный трубопровод"

Автор (авторы): Розенблат Анатолий Исаакович

Заявитель: он же

Заявка № 3451278   Приоритет изобретения 8 июня 1982г.

Зарегистрировано в Государственном реестре
изобретений СССР

15 сентября 1984г.

Действие авторского свидетельства распространяется на всю территорию Союза ССР.

Председатель Комитета

Начальник отдела

МПФ Гознака. 1979. Зак. 79-3083.

APPENDIX 12

APPENDIX 12

ROZENBLAT'S VISUAL ART WORKS (INNOVATIONS) WHICH ARE PROTECTED IN USA BY COPYRIGHT

1. Combined engine Vau 333-054 ,1995

2. Automobile Vau 333-061,1995

3. Combined rocket "AIR" Vau 342-816,1995

4. Magnetic-dynamical pipeline Vau 342-815,1995

5. Aircraft-passenger Vau 343-603,1995

6. Helicopter Vau 343-881,1995

7. Pneumatic installation Vau 345-486,1995

8. Multiple Carbide Insert and method its designing Vau 345-275 ,1995

9. Seismic Combined Nuclear System Vau 343-895,1995

10. Telescopical cutting tool   Vau 362-690 ,1996

11. Rozenblat's combined die Vau362-689,1996

12. Multi-operational die with cone Vau 362-692 ,1996

13. Break-chips insert Vau 362-691,1996

14. Isothermal method of transporting LNG and LPG by sea Vau 361-374 and Vau 360-023 ,1996

15. Analysis of chip formation in cutting stainless steels with circular segmental saws Vau 360-024,1996

16. Advanced technology for the metal-cutting processes Txu 870-753 ,1998

ABOUT THE AUTHOR

As a creative personality, independent scientist and inventor, Anatoly I. Rozenblat received a fundamental technical and art education in marine (mechanical) and manufacturing engineering, applied mathematics and patent affairs.

He completed studies at Odessa of Marine Engineers in 1967, the Polytechnic Institute 1973, Odessa Mechnikova State University in 1988 and Odessa and Moscow Patents Institutes in 1983 and 1986 respectively.

The studies provided the basis of his successful scientific and research activity, particularly in engineering and innovation. He has published more than fifty scientific articles, five books of technical literature, and has thirty inventions.

Born on 25 August 1938 in Russia and later at the beginning 1989 he immigrated to the United States. The participation in the American Society of Mechanical Engineers and Society of Naval Architects and Marine Engineers gave to him possibility to design nineteen original new technical innovations that were described by NCIO in the newsletter, *America's Inventor*, in the article entitled "Inventor from Russia Brings the New Technologies to US ".

He was invited to 26th, 27th, and 28th Israel Conference on Mechanical Engineering for presentation six scientific papers that held respectively in 1996, 1998, and 2000.

He has also received the numerous awards from International Biographical Center ((England), American Biographical Center and Marquis' *Who's Who*.

Now, he continues his professional career in Kennedy-Western University in the doctoral program in general engineering for a Ph.D. degree.

www.ingramcontent.com/pod-product-compliance
Lightning Source LLC
Chambersburg PA
CBHW081111170526
45165CB00008B/2414